super SAFE

RETAIL BEST PRACTICES AND GUIDE TO FOOD SAFETY AND SANITATION

David McSwane H.S.D. • Richard Linton, Ph.D • Nancy R. Rue, Ph.D.

Production Services and Development by
Learnovation, LLC

FMI
FOOD MARKETING INSTITUTE

Editor-in-Chief: Steve Helba
Acquisitions Editor: Vernon Anthony
Managing Editor: Mary Carnis
Production Editor: Brian Hyland
Director of Manufacturing and Production: Bruce Johnson
Design Director: Cheryl Asherman
Senior Design Coordinator: Miguel Ortiz
Editorial Assistant: Ann Brunner
Desktop Publisher: Karen J. Hall
Development Editor: Anna Graf Williams
Assistant Development Editor: David Morrow
Cover Illustration: John Wise

Pearson Education LTD.
Pearson Education Australia PTY, Limited
Pearson Education Singapore, Pte. Ltd
Pearson Education North Asia Ltd
Pearson Education Canada Ltd
Pearson Educatión de Mexico, S.A. de C.V.
Pearson Education — Japan
Pearson Education Malaysia, Pte. Ltd

This special edition is published in cooperation with Pearson Custom Publishing.

Prentice Hall

10 9 8 7 6 5 4 3 2 1

ISBN 0-536-72742-2

BA 996254

NOTE FROM FMI

Food safety has always been a paramount priority for the food industry. Today it is even more so due to improved methods of detecting safety hazards and to increased consumer concerns about both food safety and security. As the last link in the global supply chain that delivers food to consumers, retailers play a special role in protecting public health.

In recent years, the industry, government and scientific community have designed effective programs to safeguard the food supply. Some are required by law, and others are embodied in good retail practices, standard operating procedures and training that teaches lasting lessons. The programs may vary, but all feature the same basic principles, including sanitation, employee health and hygiene - all aimed to prevent contamination before customers purchase food products.

The guidelines in the following pages are one of several educational tools developed through the cooperative efforts of the Food Marketing Institute, Prentice Hall and National Computer Systems, Inc. The authors of this manual are among the world's leading food safety experts and educators. They bring decades of experience training retail food handlers and managers how to systematically prevent contamination.

This training manual identifies best practices that prevent foodborne illnesses at the retail level. Knowing these is only the beginning in fully understanding food safety. Apply this knowledge daily each time you prepare, handle, display or offer food. These practices will help ensure that your customers become loyal patrons, confident your food is safe and wholesome.

Tim Hammonds

President and CEO

Food Marketing Institute

Contents

Note from FMI	**iii**
Preface	**xix**
Acknowledgments	**xxii**
Chapter 1—Retail Food Safety	**1**
Learning Objectives	*1*
Essential Terms	*2*
New Challenges Present New Opportunities	*2*
Retail Food Establishments—Definitions	3
Food Safety—Why All the Fuss?	*5*
Why Me?	*5*
Changing Trends in Food Consumption and Choices	*6*
The Problem: Foodborne Illness	*8*
Contamination	*9*
Microorganisms (Germs or Microbes)	*11*
The Food Flow	*11*
A New Approach to an Old Problem	*12*
Facility Planning and Design	*13*
Keeping It Clean and Sanitary	*13*
Accident Prevention and Crisis Management	*14*
Education and Training Are Key to Food Safety	*15*
The Role of Government in Food Safety	*16*
The *FDA Food Code*	16
The Role of the Food Industry in Retail Food Safety	*17*
Food Protection Manager Certification	*17*
Summary	*19*
Discussion Questions	*20*
Quiz 1.1 (Multiple Choice)	*20*

Quiz 1.2 (True/False) 21

References/Suggested Readings 22

Suggested Web Sites 22

CHAPTER 2—HAZARDS TO FOOD SAFETY 23

Learning Objectives 23

Essential Terms 24

Foodborne Illness 24

Foodborne Hazards 28

Bacteria 29

Spoilage and Disease-causing Bacteria 30

Bacterial Growth 31

What Disease-causing Bacteria Need in Order to Multiply 32

Food 34

Acidity 34

Temperature 35

Time 37

Oxygen 37

Moisture 38

Potentially Hazardous Foods (PHF) 38

Ready-to-Eat Foods 40

Foodborne Illness Caused by Bacteria 41

Foodborne Illness Caused by Sporeforming Bacteria 42

Bacillus cereus 44

Clostridium perfringens 45

Clostridium botulinum 46

Foodborne Illness Caused by Non-sporeforming Bacteria 47

Campylobacter jejuni 47

Shiga toxin-producing Escherichia coli 48

Listeria monocytogenes 49

Salmonella spp. 50

Shigella spp. 51

Staphylococcus aureus 52

Vibrio spp. 53

Foodborne Illness Caused by Viruses 53
Hepatitis A 54
Norwalk virus 55
Rotavirus 55

Foodborne Illness Caused by Parasites 56
Anisakis spp. 56
Cyclospora cayentanensis 57
Cryptosporidium parvum; Giardia lamblia 58
Toxoplasma gondii 59
Trichinella spiralis 60
Problems Caused by Fungi 60

Foodborne Illness Caused by Chemicals 61
Naturally Occurring Chemicals 62
Allergens in Food 62
Ciguatoxin 64
Scombrotoxin 64
Shellfish Toxins—PSP, DSP, DAP, NSP 65
Mycotoxins 66

Added Man-Made Chemicals 67

Foodborne Illness Caused by Physical Hazards 67

Problems in Other Countries Related to Food 68

Summary 69

Discussion Questions (Short Answer) 70

Quiz 2 (Multiple Choice) 71

References/Suggested Readings 73

Suggested Web Sites 73

CHAPTER 3—PREVENTING FOODBORNE ILLNESS 75

Learning Objectives 76

Essential Terms 76

Factors That Contribute to Foodborne Illness 76

What is Time and Temperature Abuse? 78
How to Measure Food Temperatures 80
When and How to Calibrate Thermometers 83
Measuring Food Temperature 84

Preventing Temperature Abuse 87

Keep Cold Foods Cold and Hot Foods Hot! *88*

The Importance of Handwashing and Good Personal Hygiene *91*
- Outer Clothing and Apparel 96
- Personal Habits 97
- Personal Health 98

Cross Contamination *100*
- Avoiding Cross Contamination 101

Other Sources of Contamination *102*

Make Sure the Work Area is Clean and Sanitary *104*

Summary *104*

Discussion Questions (Short Answer) *105*

Quiz 3 (Multiple Choice) *105*

References/Suggested Readings *107*

Suggested Web Sites *107*

CHAPTER 4—FOLLOWING THE FOOD PRODUCT FLOW 109

Learning Objectives *109*

Essential Terms *110*

Buying from Approved Sources *110*
- Inspecting Delivery Vehicles 111

Strategies for Determining Food Quality *111*

Measuring Temperatures at Receiving and Storage *113*

Following the Flow of Food *113*
- Receiving 114
- Packaged Foods 116
- Reduced Oxygen Packaging (ROP) 116
- Red Meat Products 119
- Game Animals 121
- Poultry 121
- Eggs 123

Fluid Milk and Milk Products 124
Fluid Milk 124
Cheese 124
Butter 124
Fish 125
Fruits and Vegetables 127
Juice 128
Frozen Foods 129
Storage of Food 129
Types of Storage 130
Storage Conditions for Foods 135
Food Preparation, Handling, and Service 137
Ingredient Substitution 138
Avoiding Temperature Abuse 138
Freezing 139
Thawing 140
Cold-Holding 142
Cooking 145
Cooling 147
Hot-Holding and Reheating 150
Reduced Oxygen Packaging (ROP) 151
Food Handling 154
Discarding or Reconditioning Food 155
Refilling Returnable Containers 155
Self-Service Bar 156
Checkout and Bagging 158
Temporary Facilities and Mobile Food Facilities 159
Vending Machines 161

Home Meal Replacement *162*

Summary *163*

Discussion Questions (Short Answer) *165*

Quiz 4 (Multiple Choice) *166*

References/Suggested Readings *169*

Suggested Web Sites *170*

Chapter 5—Facilities, Equipment, and Utensils 171

Learning Objectives *171*

Essential Terms *172*

Design, Layout, and Facilities *172*

Regulatory Considerations *174*

Work Center Planning *174*

Equipment Selection *175*

Need 175

Cost 176

Size and Design 176

Construction Materials 178

Metals 178

Stainless Steel 179

Plastic 179

Wood 180

Types of Equipment *180*

Cooking Equipment 181

Ovens 181

Refrigeration and Low-Temperature Storage Equipment 185

Reach-in Refrigeration 187

Refrigerated Display Cases 188

Walk-in Refrigerators 188

Cook-Chill and Rapid-Chill Systems 189

Hot-Holding Equipment 190

Other Types of Food Equipment 191

Slicers 191

Mixers, Grinders, Choppers, Tenderizers and Saws 192

Ice Machines 194

Ambient Temperature Display Equipment 196

Live Seafood Display and Holding Tanks 197

Single-Service and Single-Use Articles 198

Warewashing Equipment 199

Manual Warewashing 200

Mechanical Warewashing 201

Installation *203*

Maintenance and Replacement *203*

Lighting 204

Heating, Ventilation, and Air Conditioning (HVAC) 206

Summary 208

Discussion Questions (Short Answer) 209

Quiz 5 (Multiple Choice) 210

References/Suggested Readings 212

Suggested Web Sites 213

CHAPTER 6—CLEANING AND SANITIZING OPERATIONS 215

Learning Objectives 216

Essential Terms 216

Principles of Cleaning and Sanitizing 216

Removal of Food Particles 217

Application of Cleaning Agents 218

Soaking 218

Spray Methods 218

Clean-in-Place Systems 219

Abrasive Cleaning 219

Rinsing 220

Factors Affecting Cleaning Efficiency 220

Type of Soil to Be Removed 220

Water Quality 220

Detergents and Cleaners to Be Used 221

Water Temperature 224

Velocity or Force 224

Amount of Time the Detergent or Cleaner Remains in Contact with the Surface 224

Concentration of the Detergent or Cleaner 224

Cleaning Frequency 224

Sanitizing Principles 227

Heat Sanitizing 227

Chemical Sanitizing 230

Factors that Affect the Action of Chemical Sanitizers 231

Chlorine 233

Iodine 234

Quaternary Ammonium Compounds (quats) 234

Mechanical Warewashing 235
Mechanical Warewashing Process 236
Manual Warewashing 240
Cleaning Fixed Equipment 243
Cleaning Environmental Areas 249
Ceilings 250
Walls 250
Floors 250
Equipment and Supplies Used for Cleaning 251
Summary *254*
Discussion Questions (Short Answer) *255*
Quiz 6 (Multiple Choice) *256*
References/Suggested Readings *258*
Suggested Web Sites *258*

CHAPTER 7—ENVIRONMENTAL SANITATION AND MAINTENANCE **261**

Learning Objectives *262*
Essential Terms *262*
Condition of the Establishment *263*
Proper Water Supply and Sewage Disposal System 263
Condition of Building *264*
Floors, Walls, and Ceilings 265
Floors 266
Walls and Ceilings 267
Restroom Sanitation 268
Handwashing Facilities 269
Plumbing Hazards in Retail Food Establishments *271*
Cross Connections 271
Backflow 272
Methods and Devices to Prevent Backflow 272
Backflow Prevention Devices on Carbonators 276
Grease Traps 276
Garbage and Refuse Sanitation *277*
Inside Storage 278
Outside Storage 279

Pest Control *281*
Insects 282
Flies 282
Cockroaches 285
Moths and Beetles 287
Rodents 289
Signs of Rodent Infestation 290
Rodent Control 291
Integrated Pest Management (IPM) 293
Summary *294*
Discussion Questions (Short Answer) *295*
Quiz 7 (Multiple Choice) *296*
Suggested Reading/References *298*
Suggested Web Sites *299*

CHAPTER 8—ACCIDENT PREVENTION AND CRISIS MANAGEMENT **301**

Learning Objectives *301*
Essential Term *302*
Safety in Retail Food Establishments *302*
What to Do if an Accident Occurs 304
Common Types of Injuries 304
Falls 304
Cuts 305
Burns 307
Poisoning 307
Back Injuries 308
Blood and Body Fluid Borne Diseases 308
Employee Medications 309
Cardiopulmonary Resuscitation (CPR) and First Aid for Choking 309
Self-Inspection Safety Checks 309
Incident Reports 311
Fire Safety 311
Fire Extinguishers 311
Hood Systems 312
Sprinkler Systems 313

Operating a Forklift 313

Crisis Management *313*

Water Supply Emergency Procedures 314

Bioterrorism 315

Foodborne Illness Incident or Outbreak *317*

Check the Product Source 317

Formal Investigations 317

Summary *318*

Quiz 8 (Multiple Choice) *319*

Reference/Suggested Readings *320*

Suggested Web Sites *320*

CHAPTER 9—EDUCATION AND TRAINING **321**

Learning Objectives *322*

Essential Terms *322*

Market Need *322*

Learners *323*

Current Information and Technical Challenges *324*

Training is Important *325*

Barriers to Training *326*

The Agenda for Training in Retail Food Establishments *326*

Needs Assessments *327*

Best Practices Recommended Times for Training *328*

Standard Operating Procedures Federal, State, and Local Rules and Regulations *329*

Hazard Analysis Critical Control Point System *329*

Environment *329*

Conduct the Training *330*

Technology *331*

Evaluation *331*

Training Moments *332*

Summary 332

Quiz 9 (Multiple Choice) 333

References/Suggested Readings 334

Suggested Web Sites 334

CHAPTER 10—FOOD SAFETY MANAGEMENT PROGRAMS 337

Learning Objectives 338

Essential Terms 338

The Problem 338

The Solution 339

Start with a Good Set of Instructions 340

The HACCP Approach to Food Safety 343

The Seven Principles in a HACCP System 344

Retail HACCP: A Process Approach 346

Food Safety Management and HACCP Requirements 352

Good Retail Practices 352

Food Recall Procedures 355

Coding and Product Identification 355

Challenges of Food Safety Management 356

Summary 358

Discussion Questions (Short Answer) 358

Quiz 10 (Multiple Choice) 359

References/Suggested Readings 360

Suggested Web Sites 361

CHAPTER 11—FOOD SAFETY REGULATIONS 363

Learning Objectives 363

Essential Terms 364

State and Local Regulations 364

Permit to Operate 365

Federal Agencies *367*
Food and Drug Administration (FDA) 368
U.S. Department of Agriculture (USDA) 370
U.S. Department of Commerce (USDC) 371
Environmental Protection Agency (EPA) 372
Centers for Disease Control and Prevention (CDC) 372
Occupational Safety and Health Administration (OSHA) 373
Federal Trade Commission (FTC) 373
Other National Food Safety Related Organizations 373

Food Law *374*
Federal Food, Drug, and Cosmetic (FD&C) Act 374
1906–Federal Meat Inspection Act 375
1957–Federal Poultry Products Inspection Act 375
1970–Egg Products Inspection Act 375
1996–Food Quality Protection Act 375

Food Labeling *376*
Nutrition and Ingredient Labeling 376
Safe Food-Handling Label 377
Product Dating Labels 378

Recent Initiatives in Food Safety *378*
National Food Safety Initiative 378
FoodNet 379
PulseNet 379
The Fight BAC!TM Campaign 380
Consumer Advisory 380

Summary *381*

Quiz 11 (Multiple Choice) *382*

References/Suggested Reading *384*

Suggested Web Sites *385*

APPENDIX A — ANSWERS TO CASE STUDIES AND QUIZZES **387**

APPENDIX B — SUMMARY OF AGENTS THAT CAUSE FOODBORNE ILLNESS **393**

APPENDIX C — EMPLOYEE HEALTH – DISEASE OR MEDICAL CONDITION REPORTABLE CONDITIONS AND ACTIVITIES **399**

Exclusions and Restrictions *400*

Removal of Exclusions and Restrictions *401*

APPENDIX D — CONVERSION TABLE FOR FAHRENHEIT AND CELSIUS FOR COMMON TEMPERATURES USED IN FOOD ESTABLISHMENTS **403**

APPENDIX E — SPECIFIC ELEMENTS OF KNOWLEDGE EVERY RETAIL FOOD PROTECTION MANAGER SHOULD KNOW **405**

GLOSSARY **411**

INDEX **431**

Preface

Food safety has become an issue of special importance for the retail food industry. There are many opportunities for food to be contaminated between production and consumption. Food can be contaminated at the farm, ranch, orchard, or in the sea. Food also can be contaminated at food processing plants and during transport to retail food establishments. Finally, food can be contaminated during the last stages of production, at retail establishments, and by patrons in their homes.

Food safety is especially critical in retail food establishments because this may be the last opportunity to control or eliminate the hazards that might contaminate food and cause foodborne illnesses. Even when purchased from inspected and approved sources, ingredients may contain contaminants when they arrive at a retail food establishment. It is important to know how to handle ingredients safely and how to prepare food in a manner that reduces the risk of contaminated food being sold to your patrons.

The Food and Drug Administration (FDA) is one of the federal agencies responsible for protecting our nation's food supply. The Agency recognizes the importance of food safety in retail food establishments. In fact, the agency's 2001 Food Code recommends that retail food managers be able to demonstrate knowledge in food safety. This knowledge will be very helpful to managers as they create and implement food safety management programs within their operations.

Food safety in retail food establishments begins with managers who are knowledgeable about food hazards and who are committed to implementing proper food-handling practices in their facility. It also requires properly trained food workers who understand the essentials of food safety and sanitation and who will not take short cuts when it comes to food safety.

The authors of this textbook have been training retail food managers and employees for over 25 years. Many excellent resources are available for this type of training. However, the authors wanted to create a book that was customized for the retail food industry and would meet the training needs of supermarkets, superstores, food warehouses, limited assortment stores, convenience stores, food warehouses, and other types of traditional and nontraditional stores.

The Best Practices materials have been proven effective for teaching food safety and sanitation to many different audiences. The authors recommend the textbook and supplemental materials in-class and home study courses to prepare retail food establishment managers to take a national food protection manager certification examination.

One of the most important tasks you face is to train and supervise food workers. Your knowledge is useless if you do not teach employees the correct way to handle food. Learn to recognize any break in standard operating procedures that might endanger food safety. Avoid the problems related to embarrassment, loss of reputation, and financial harm that accompany a foodborne disease outbreak.

Some state and local jurisdictions have passed legislation to require certification of one or more managers in each retail food establishment. Other jurisdictions are considering doing the same. Most certification programs require candidates to pass a written examination to demonstrate knowledge of food safety and sanitation principles and practices. Some jurisdictions require completion of a food safety course before taking the exam.

There is growing support for a nationally recognized examination and credential for retail food protection managers. The Conference for Food Protection (CFP) recognizes food protection manager examinations from several providers. Contact the CFP at www.foodprotect.org for more information about the test recognition process and the various providers who have one or more forms of their examination recognized by the Conference.

The team of McSwane, Linton, and Rue wanted to produce a book to accompany the *FDA's 2001 Food Code*. We trust that you will find this text accurate, comprehensive, and, most of all, useful.

David Z. McSwane, H.S.D., REHS, CFSP, is an Associate Professor of Public and Environmental Affairs at Indiana University. He has over 25 years of experience in food safety and sanitation working in state and local regulatory agencies and as a consultant to the food industry. Dr. McSwane has published numerous articles and presented papers on a variety of subjects related to food safety.

Dr. McSwane is a nationally recognized trainer in food safety and sanitation. He has taught courses at the university level and for regulatory agencies, retail food establishments, food industry trade associations, vocational schools, and environmental health associations throughout the United States. Dr. McSwane is a recipient of the Walter S. Mangold award. This is the highest honor bestowed by the National Environmental Health Association. He has been a correspondent for the Food Protection Report and is a member of the Environmental and Public Health Council at Underwriters Laboratories Inc.

Richard Linton, Ph.D., is Professor of Food Safety at Purdue University. His expertise is in the development and implementation of food safety and food quality programs, specifically in Hazard Analysis Critical Control Point (HACCP) systems. Dr. Linton spent twelve years working in retail food establishments. In recent years, he has educated food workers and managers from all segments of the industry throughout the nation and the world.

Dr. Linton has developed several different types of food safety training programs for retail food managers and workers. He also works closely with the retail food industry on research projects that help improve the quality and safety of the food they serve.

Nancy Roberts Rue, Ph.D., R.N., has a background in teaching in technical education. Her doctorate is in educational leadership and curriculum and instruction from the University of Florida. With thirty years of experience in higher education and evaluation at Indiana University and St. Petersburg Junior College, she is dedicated to the task of building educational materials that meet the needs of those who want to learn. She is now an independent writer and consultant on training and development techniques.

There are several reference books available in the field of food safety. However, most of these resources are designed for restaurants, institutional kitchens, and other types of food establishments. The authors of Retail Best Practices and Guide to Food Safety and Sanitation believe there is need for a text that applies food safety principles to retail food establishments such as supermarkets and convenience stores.

The authors wish all readers success in their food safety and sanitation activities. Regardless of where you work, you must always remember—foodborne illness is a preventable disease. Follow the basic rules of food safety and you will enjoy a satisfying career in the retail food industry.

ACKNOWLEDGMENTS

The authors wish to thank their families and colleagues who, through their support and patience, helped us with the first edition of this book. John Wise is a tremendous asset to this project. His drawings enable us to present information that is technical and sometimes tedious in a way that make it easy to understand. Karen J. Hall, David Morrow, and George Williams from Learnovation, desktop publishing and developmental editors, gave us the technical support needed to make the book come alive.

Special thanks goes to Tom Compbell and Vincent P. Walter, the project photographers for their innovative and inspired services. Scott Alkinburgh of Marsh Supermarkets and Don Loutzenhiser of Martin's Supermarkets each graciously shared their knowledge and facilities with us during the writing of the book. A heartfelt acknowledgement goes to Vernon Anthony, the acquisitions editor at Prentice Hall, for his vision and support during this process.

A grateful thanks to goes to Timothy Weigner and Laurie Williams, Food Safety Programs at Food Marketing Institute, for their technical expertise and insight into supermarket food safety programs.

We would like to acknowledge and recognize the supermarket leaders who provided expertise and review of the program as volunteers. They include:

Ruth Yong
The Kroger Co.

Cory Hedman
Hannaford Bros. Co.

Robert Brown
Shaw's Supermarkets, Inc.

Dave Pettigrew
Spartan Stores, Inc.

Monica Martin
Farm Fresh, Inc.

Garey Walker
Giant Food, Inc.

Scott Alkinburg
Marsh Supermarkets

John Dammon
The Copps Corporation

Krag Swatz
Lund Food Holding, Inc.

Michael Hillyer
Wal Mart Stores, Inc.

Don Loutzenhizer
Martin's Super Markets

George Sherback
The Stop & Shop Supermarket

Gillian Kelleher
Wegmans Food Markets, Inc.

Michael Magner
Price Chopper Supermarkets

Cas Tryba
Big Y Foods, Inc.

1 RETAIL FOOD SAFETY

Bad Burger Spoils Fourth of July Holiday

It was the Fourth of July weekend in a popular lake resort town. This was expected to be one of the busiest weekends of the year. The shelves in the retail food establishments were heavily stocked. One of the most popular establishments was using special sales to attract customers for holiday shopping. Ground beef was one of the featured items in the meat department.

During the holiday period, the retail food establishment manager received a telephone call from the local health department. Apparently, the ground beef sold at this store was suspected of being contaminated with Shiga toxin-producing Escherichia coli *bacteria. At the time of the phone call, 19 people had contacted the health department complaining of nausea, vomiting, and bloody diarrhea. All of them had purchased ground beef at this store in the last few days. The health inspector told the manager a foodborne disease outbreak was suspected. A foodborne disease outbreak occurs when two or more people become ill after eating a common food. The health department was conducting an investigation to find the cause of the illness.*

Several hundred pounds of ground beef were sold in the past few days. How did these bacteria contaminate the ground beef? What can the retail food establishment manager do to prevent others from getting sick?

Learning Objectives

After reading this chapter, you should be able to:

- ▲ Recognize the importance of food safety and sanitation as the basis for preventing foodborne illness in retail food establishments.
- ▲ Identify various causes of foodborne illness in retail food establishments.

- ▲ List methods to prevent or reduce foodborne illness in retail food establishments.
- ▲ Recognize the role of the *FDA Food Code* as a source of recommendations for safe food handling.
- ▲ Recognize the importance of food safety education and training.
- ▲ Explain the role of government regulation (federal, state, and local) in retail food safety.
- ▲ Identify the term Hazard Analysis Critical Control Point (HACCP) as a system for food protection.
- ▲ Recognize the need for food protection manager certification.

Essential Terms

Cleaning	**Foodborne illness**
Contamination	**Microorganisms**
Convenience store	**Germs**
FDA Food Code	**Microbes**
Food establishments	**Retail food establishment**
Foodborne disease outbreak	**Sanitary**

New Challenges Present New Opportunities

The food industry is one of America's largest enterprises. It employs about one-quarter of the nation's work force and produces 20% of America's Gross Domestic Product (GDP). Billions of dollars worth of food are sold each year. Americans have made food a prominent part of their business and recreational activities.

The food industry is made up of businesses that produce, manufacture, transport, and distribute food for people in the United States and throughout the world. Food production involves many activities that occur on farms and ranches, in orchards, and in fishing operations. Food manufacturing takes the raw materials harvested by producers and converts them into forms suitable for distribution and sale. The retail distribution system consists of the many food operations that store, prepare, package, serve, display, vend, or otherwise provide food for human consumption. The term **food establishment** refers to all facilities involved in food distribution. A list of food establishments is presented in Figure 1.1.

Figure 1.1 Types of Food Establishments

Retail Food Establishments—Definitions

During 2000, U. S. grocery sales grew by 3.4%, reaching $570 billion. Retail channels have changed dramatically over the last few years (Willard Bishop Consulting, 2001). There are 15 different store formats currently operating in the retail food industry. This list includes:

- ▲ Conventional Supermarket–The original supermarket format offering a full line of groceries, meat, and produce with at least $2 million in annual sales.
- ▲ Superstore–A larger version of the conventional supermarket with at least 40,000 square feet in total selling area. Superstores offer an expanded selection of non-foods (at least 10% GM/HBC).
- ▲ Food/Drug Combo–Combination of superstore and drug store.
- ▲ Warehouse Store–A low margin grocery store offering reduced variety, minimal décor, and streamlined merchandising.

- ▲ Limited Assortment Stores–A barebones, low-priced grocery that provides limited services and carries fewer than 2,000 items with limited perishable items.
- ▲ Other–The small corner grocery store that carries a limited selection of staples and convenience goods.
- ▲ Convenience Store (Traditional)–Small, higher margin store that offers a selection of staples, non-food, and other convenience foods.
- ▲ Convenience Store (Petroleum-based)–Primarily gasoline stations with a convenience store.
- ▲ Nontraditional Grocery:
  - ▼ Hypermarket
  - ▼ Wholesale Club
  - ▼ Mini-club
  - ▼ Supercenters
  - ▼ Deep Discount Drug Store
  - ▼ Internet.

The size and scope of the retail food establishment industry in the United States is immense. According to the Food Marketing Institute, there are more than 127,000 retail food establishments with an average of 49,225 items in their inventories. The retail food establishment industry employs more than 3.5 million people. The size and diversity of the food industry has made it virtually impossible for regulatory officials to continuously monitor all aspects of food safety. Therefore, modern food safety programs must involve governmental agencies and the food industry working closely together to ensure the safety of our food supply.

Retail food establishments and the customers they serve are very different. Some establishments will prepare foods while others will sell only prepackaged, ready-to-eat foods. As you study problems related to food safety, keep this diversity in mind.

This book focuses on information related to food safety and sanitation in any retail food establishment. The term **retail food establishment** includes any store selling a line of dry retail food, canned goods or non-food items plus perishable items. A convenience store is a compact, drive-to store offering a limited number of high-convenience items. The information would also apply to other types of retail food stores such as supercenters, warehouse stores, and wholesale clubs.

Food Safety—Why All the Fuss?

Everyone knows the United States has one of the safest food supplies in the world. So why all the fuss? The answer is simple. Foodborne illness happens, and it adversely affects the health of millions of Americans every year. **Foodborne illness** is the sickness some people experience when they eat contaminated food. It impairs performance and causes discomfort. Estimates of the number of cases of foodborne illnesses vary greatly. A recent report issued by the Centers for Disease Control and Prevention (CDC) estimates foodborne diseases cause approximately 76 million illnesses, 325,000 hospitalizations, and 5,000 deaths in the United States each year (Mead et. al., 1999).

Foodborne illness also has a major economic impact. It costs our society billions of dollars each year. These costs occur in the form of medical expenses, lost work, and reduced productivity by victims of the illness, legal fees, punitive damages, increased insurance premiums, lost business, and loss of reputation. Consumer confidence goes down when a foodborne disease outbreak is reported. Retail food establishments that have been linked to a foodborne illness are likely to receive more frequent inspections by regulatory agencies. The establishment can also be closed if it is determined that is the best way to protect public health. A foodborne disease outbreak will have a negative impact on the owners, managers, and employees of the retail food establishment where it occurred and on the rest of the food industry as well.

The Food Marketing Institute (FMI) reports nearly 75% of shoppers are completely or mostly confident the food they buy in their retail food establishment is safe. Seven out of every 10 shoppers say product safety is "very important" to them when they shop for food. That is a valid reason to keep the food sold in your operation as safe as possible.

Why Me?

You may be asking yourself, "What does all of this have to do with me?" The answer is "PLENTY." Customer opinion surveys show cleanliness, price and food quality are the top three reasons people use when choosing a place to eat and shop for food. Customers expect their food to taste good and not make them sick. It is the responsibility of every retail food establishment owner, manager, and employee to prepare and serve safe and wholesome food and preserve their clients' confidence.

The CDC reports in most foodborne illness outbreaks the mishandling occurred within retail food operations (restaurants, retail food establishments, schools, churches, camps, institutions, and vending locations) where foods are prepared, served, and sold to the public. These foods may be eaten at the food establishment or sold for preparation and consumption elsewhere.

The nature of food and the extent to which it is handled make the opportunity for contamination commonplace. Most cases of foodborne illness in retail food establishments are caused by foods that have been:

- ▲ Improperly cooked and/or held at improper temperatures
- ▲ Handled by infected food employees who practice poor personal hygiene
- ▲ Exposed to disease-causing agents by cross contamination
- ▲ In contact with contaminated equipment not properly cleaned and sanitized
- ▲ Obtained from unsafe sources.

Each of these factors will be discussed in more detail in later chapters of this book.

As a retail food manager, you must understand foodborne illness can be prevented if the basic rules of food safety are routinely followed. Figure 1.2 shows some of the key elements of food safety. Prevention of foodborne illness must be a goal in every retail food establishment.

Changing Trends in Food Consumption and Choices

People are buying different foods today, and the retail food industry is changing to meet the needs of the public. For example, the consumption of fresh produce is on the rise. We know fruits and vegetables can provide many health benefits. Convenience foods are also more popular. Many of these foods are ready-to-eat with little or no preparation required. Retail food establishments also involve more food preparation than ever before. Salad bars, pizza stands, Chinese food, sushi, and bakeries are right on the premises.

Due to changes in our eating habits and more knowledge about food safety hazards, recommendations for safe food handling are always changing. For example, we used to recommend cooking ground beef to an internal temperature of 140°F (60°C) to destroy dangerous bacteria that might be present. But that was before Shiga toxin-producing *Escherichia coli* bacteria came on the scene. Now retail food establishments are required to cook ground beef to an internal temperature of 155°F (68°C) for 15 seconds. This higher temperature is needed to destroy the Shiga toxin-producing *Escherichia coli* bacteria that may be present in the raw meat. The safety of apple cider did not use to be a concern. However, Shiga toxin-producing *Escherichia coli* bacteria have changed that. Now apple cider must be pasteurized in order to assure it is safe to drink. Containers of unpasteurized cider must have a label that advises customers of the potential health risk associated with drinking this product.

Figure 1.2 Elements of Safe Food

Food safety issues and concerns change constantly. Some of the key food safety concerns today include *Listeria monocytogenes* in ready-to-eat processed foods, Shiga toxin-producing *Escherichia coli* in raw meat and unprocessed fruit juices, Hepatitis A virus in deli sandwich operations and shellfish, and parasites in fresh produce. The future may bring other types of problems. Technologies used in the food industry are also changing. A lot of research is being done on irradiation and other types of food-processing methods that do not use heat. They may become more common in the future. You need to keep up with this information as it relates to retail food establishment operations.

Customers have less time to prepare food because more of them are working outside the home. As a result, they are buying more ready-to-eat food or products that require minimal preparation in the home. More than half the food eaten by Americans is prepared in food processing plants, restaurants, retail food establishments, delicatessens, cafeterias, institutions, and other sites outside the home. These foods are produced using a variety of processing, holding, and serving methods that help protect them from contamination.

The Problem: Foodborne Illness

Foodborne illness is a disease caused by the consumption of contaminated food. A **foodborne disease outbreak** is defined as an incident in which two or more people experience a similar illness after eating a common food.

Recent outbreaks of foodborne illness have been caused by:

- ▲ Shiga toxin-producing *Escherichia coli* bacteria in lettuce, unpasteurized apple cider, and radish sprouts
- ▲ *Salmonella spp.* in cut melons, alfalfa sprouts, ice cream, and dry cereal
- ▲ Hepatitis A virus in raw and lightly cooked oysters
- ▲ *Listeria monocytogenes* in hot dogs and luncheon meats.

A growing number of people are especially susceptible to foodborne illness. Examples of highly susceptible populations include:

- ▲ The very young
- ▲ The elderly
- ▲ Pregnant or lactating women
- ▲ People with impaired immune systems due to cancer, AIDS, HIV, diabetes, or medications that suppress response to infection.

Foodborne illness can cause severe reactions, even death, for individuals in these high-risk categories. The availability of a safe food supply is critical to these people.

Outbreaks of foodborne illness receive widespread media attention. Newspapers, television, and radio have become sensitive to issues surrounding food safety. As a result, public awareness about the problem, and how it occurs, has increased dramatically in the last few years.

Contamination

Whether a food item is prepared "from scratch" or arrives in a ready-to-eat form, there are opportunities for it to become contaminated before it is consumed. **Contamination** is the presence of substances or conditions in the food that can be harmful to humans. Bacteria and viruses pose the greatest food safety challenges for retail food establishments.

Foods can become contaminated at a variety of points as the food flows from the farm to the table (Figure 1.3). Raw foods can be contaminated at the farm, ranch, or on board a commercial fishing boat. Contamination can also occur as foods are handled during processing and distribution. Measures to prevent and control contamination must begin when food is harvested and continue until the food is consumed.

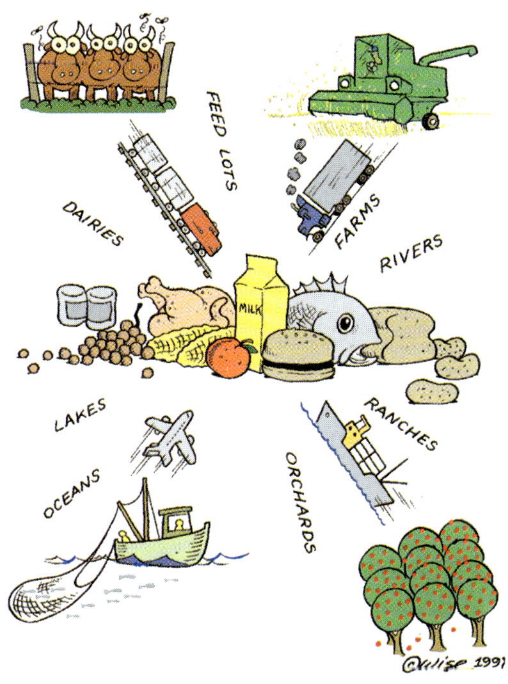

Figure 1.3 Foods Can Become Contaminated at Several Points Between the Farm and the Table

Figure 1.4 shows some of the more common sources of contamination such as soil, water, air, plants, animals, and humans. Contaminants present an "invisible challenge" because you can't see them with the naked eye. Many types of food contamination can cause illness without changing the appearance, odor, or taste of food.

Figure 1.4 Sources of Contamination

Germs can be transferred from one food item to another by cross contamination. This typically happens when microbes from a raw food are transferred to a cooked or ready-to-eat food by contaminated hands, equipment, or utensils.

Microorganisms (Germs or Microbes)

Microorganisms (also called germs or microbes) are the most common types of food contamination. Microorganisms include bacteria, viruses, parasites, and fungi that are so small they can only be seen with the aid of a microscope. Microbes are everywhere around us—in soil, water, air, and in and on plants and animals (including humans). The microorganisms that cause most foodborne illnesses are bacteria and viruses. These organisms get their energy from the food in which they live and reproduce—often the same foods we eat.

Most microorganisms are harmless. However, some microbes can cause problems when they get into food. The microbes that must be controlled in a retail food establishment are the ones that cause foodborne illness and food spoilage. Figure 1.5 shows three types of bacteria that cause foodborne illnesses. It is important to remember the germs that cause foodborne illnesses usually do not alter the taste, odor, and appearance of the food.

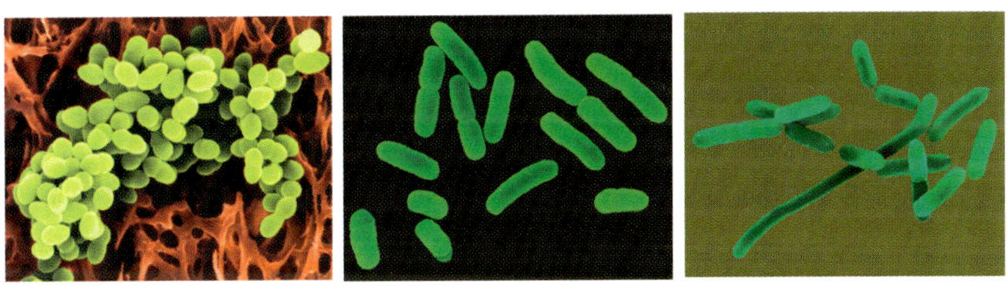

a. ***Staphylococcus aureus*** b. **Shiga toxin-producing *Escherichia coli*** c. ***Listeria Monocytogenes***

Figure 1.5 Common Causes of Foodborne Illness

(Copyright Dennis Kunkel Microscopy, Inc.)

The Food Flow

Food flow consists of food products, and the ingredients used to make them, as they flow through a retail food establishment. The food flow begins with the purchase of safe and wholesome ingredients from approved sources. Once the food is delivered, it then flows through receiving into storage. The final stages in the food flow are preparation, display, and sale. Preparation and display include all the activities that occur between storage and consumption of the food by your customers.

Preparation steps frequently involve thawing, cooking, cooling, reheating, hot-holding, cold-holding, handling, and display. Improper food handling and display can lead to foodborne illness. For potentially hazardous foods (those that support bacterial growth), time and temperature must be monitored and controlled. Handling of ready-to-eat foods (those not requiring cooking) can also involve a lot of contact with an employee's hands or with food-contact surfaces. It is always important for food employees to practice good personal hygiene and use proper handwashing techniques. However, when handling and displaying ready-to-eat foods, it is crucial!

Since handling and display are typically the last steps prior to consumption of the food, efficient monitoring and control of safe food practices are also critical. As products flow through a retail food establishment, they are handled many times and in many different ways. Controlling food safety and quality is of utmost importance. You must learn to recognize foods that could have been contaminated prior to delivery and understand what is required to keep these products safe until you sell or serve them to your customers.

A New Approach to an Old Problem

Food industry professionals and regulatory officials agree better ways to protect people from foodborne illness must be found. The *FDA Food Code* recommends using the Hazard Analysis Critical Control Point (HACCP) system as an important component of your food safety management program. This innovative system was developed by the Pillsbury Company in the 1960s for the NASA space program. Officials in the space program realized a foodborne illness in space could present a life-threatening situation. Therefore, it was critical all food used by the astronauts be as near 100% safe as possible.

The HACCP system follows the flow of food through the retail food establishment and identifies each step in the process where contamination might cause the food to become unsafe. When a problem step is identified, action is taken to make the product safe or, if that is not feasible, the food is discarded. Retail food stores can control the flow of food by setting up a food safety management program that includes a HACCP system. The HACCP system should be designed to accommodate the types of products served and the production equipment and processes used in the establishment. Records used in conjunction with the HACCP system are kept on file at the retail food establishment. These records will be periodically reviewed by managers and regulatory inspectors to ensure the system is working properly.

HACCP systems for monitoring food safety have been used in food manufacturing plants for many years. Now this system is being recommended for use in retail food establishments as well. A more detailed discussion of the HACCP system as a component of your total food safety management program is presented in Chapter 10 of this book.

Facility Planning and Design

A well-planned facility with a suitable layout is essential for the smooth operation of any retail food establishment. Layout, design, and facilities planning directly influence:

- ▲ Employee safety and productivity
- ▲ Labor and energy costs
- ▲ Customer satisfaction.

Some retail food establishments are located in buildings designed and constructed specifically for a food operation. Others are located in buildings converted to accommodate food preparation, handling, display and sales. Either way, the better your facility is planned, the easier it will be to achieve your food safety goals and earn a profit.

Keeping It Clean and Sanitary

Customers' impressions about the cleanliness of an operation are frequently influenced by what they see inside and outside the establishment. The facility must be clean and free of litter. Shopper satisfaction is highest in food establishments that are clean and bright and where quality food products are safely handled and displayed. By increasing customer satisfaction, you will develop a devoted clientele who will regularly patronize your establishment.

It is the responsibility of every person working in the food industry to keep things clean and sanitary. Effective cleaning of equipment reduces the chances of food contamination during preparation, storage, display, and service (see Figure 1.6). **Cleaning** involves removal of visible soil from the surfaces of equipment and utensils. **Sanitary** means healthful or hygienic. It involves reducing the number of disease-causing microorganisms on the surface of equipment and utensils to acceptable public health levels. Something that is sanitary poses little or no risk to human health. Good sanitation also minimizes attraction of pests, increases the length of time equipment is in service, improves employee morale and efficiency, and is important from other aesthetic considerations.

Poor performance by the cleanup crew may result in dirty utensils and equipment that could harm customers and hinder business. Proper sanitation practices must be employed at all times to protect the health and well-being of customers.

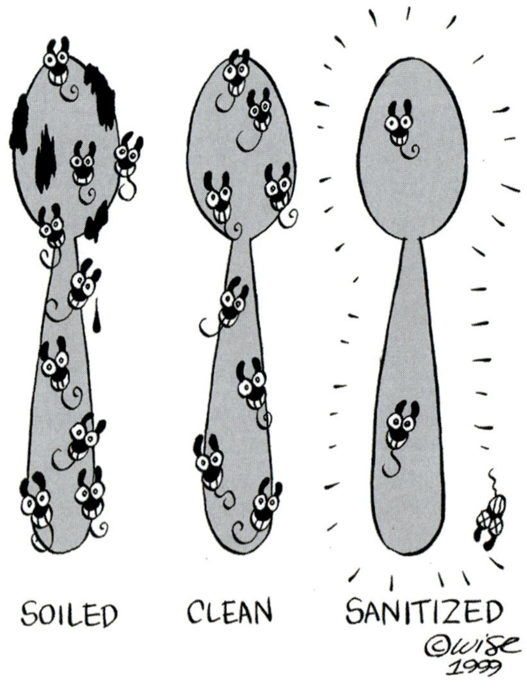

Figure 1.6 Sanitary Makes Sure It's Safe

Accident Prevention and Crisis Management

Accident prevention programs are necessary in every retail food establishment. The cost of accidents may mean the difference between profit and loss. Beyond financial responsibilities, the loss of an employee's skills can disrupt operations and cause additional stress on other employees. Ensuring a safe environment for employees and customers requires continuous monitoring, but the rewards make it worth the effort.

Human error always will be a factor in retail food establishments, but training and proper equipment can help employees avoid accidents. Oven doors left open on deck ovens, poorly stacked boxes in storage areas, spilled food, and many other situations can lead to serious injuries. For example, a fall on a wet floor can cause expenses running into the thousands of dollars for lost time, employee's compensation, and medical care.

In retail food establishments, another kind of crisis occurs when water supplies, electricity, or sewer systems are disrupted. The *FDA Food Code* contains specific instructions on how to operate when basic services to your establishment are lost. Public health departments can also offer assistance when you are not sure how to proceed.

Storms, tornados, hurricanes, or floods cause conditions that require changes to maintain food safety. A good disaster plan is invaluable in times of need. Managers and supervisors are expected to know how to handle these and other emergencies when they occur.

Education and Training Are Key to Food Safety

The prevention of foodborne illness begins with the knowledge of where contaminants come from, how they get into food, and what can be done to control or eliminate them. You are not expected to become a food microbiologist, but a working knowledge of what causes foodborne illness is valuable information. You and your employees must know the correct way to manage food safety and sanitation.

The importance of teaching employees about food safety is increased by the global nature of our food supply. Control of factors during growth, harvest, and shipping is not always possible when food is produced in so many different parts of the world. Also, an error in time and temperature management, cross contamination, or personal health and hygiene of food employees can cost a life among our vulnerable citizens. As you have learned, storage, preparation, holding, display and handling procedures are critical in the prevention of foodborne illness. Employees do not typically come to the job knowing this information. They have to be trained.

Managing the safety of food involves controlling the supply, maintaining sanitary facilities, and training employees to work with food safely. Knowledge, skills, and abilities must focus on food safety practices that will protect the public from foodborne illness. Food safety programs require food managers and food employees to have the knowledge and skills needed to keep food safe. It is not enough the manager knows about food safety. Employees must also know how to perform correctly.

The Role of Government in Food Safety

The purpose of government regulation in food safety is to oversee the food-producing system and protect food intended for human consumption. Governmental agencies enforce laws and rules to protect food against adulteration and contamination. Regulatory personnel monitor both the process and the product to ensure the safety of the food we eat.

Federal, state, and local regulators inspect food products and production facilities as a way of detecting problems. However, there are not enough inspectors and other resources to adequately administer the program. The enormous size of the food industry makes it virtually impossible for regulatory inspectors to monitor effectively all aspects of food safety. Therefore, food safety programs involve governmental agencies and the food industry working together to ensure the safety of our food supply.

Either the U.S. Food and Drug Administration (FDA) or the U.S. Department of Agriculture (USDA) oversees foods manufactured and distributed throughout the United States. Most retail food establishments are governed by state or local inspection services delivered by the health department or the Department of Agriculture. Every state has its own laws governing food safety and sanitation for retail food establishments. You are expected to know the policies and procedures in your state.

The FDA Food Code

In 1993, the U.S. Food and Drug Administration (FDA) published the Food Code which replaced three earlier model codes–The Food Service Sanitation Manual Including a Model Food Service Sanitation Ordinance of 1976, The Vending of Food and Beverages Including a Model Sanitation Ordinance of 1978, and The Retail Food Store

Sanitation Code of 1982–with one code. The FDA has made a commitment to revise the Food Code every two years, with the assistance of experts from state and local government, industry, professional associations, and colleges and universities. It was determined a biannual revision was necessary to keep pace with changes occurring in food safety and the food industry. In this text, the 2001 Food Code will serve as the resource from which safe food-handling recommendations are taken and will be referred to as the ***FDA Food Code*** from this point forward.

The *FDA Food Code* is not a law. Rather, it is a set of recommendations designed to be used as a model by state and local jurisdictions when formulating their own rules and regulations. You will need to obtain a copy of the health rules and regulations that apply to the food establishments in your jurisdiction.

The Role of the Food Industry in Retail Food Safety

The retail food industry is assuming greater responsibility for overseeing the safety of its own processes and products. Customers expect and deserve food safe to eat. If a retail food establishment is involved in a foodborne disease outbreak, customers may retaliate by taking their business elsewhere or by seeking legal action. Financial loss and damaged reputation are some of the outcomes of a foodborne disease outbreak that can cause serious harm to the establishment found responsible for the problem. One means of preventing the harmful effects of a foodborne disease outbreak is to start a food safety management program in the food establishment. This helps ensure proper safeguards are used during food production, handling, and display. The ability to prove a food safety system was in place at the time a foodborne disease outbreak occurred is very important. It has been deemed an acceptable defense in court cases where victims of foodborne illness have sought punitive damages.

Food Protection Manager Certification

Several organizations including the Conference for Food Protection (CFP), the FDA, and the trade associations for food retailers, such as the Food Marketing Institute (FMI), have recommended food manager certification as a means of ensuring individuals who are in charge of retail food establishments are knowledgeable about food safety. A certified food protection manager is a person who:

- ▲ Is responsible for identifying hazards in the day-to-day operation of a food establishment that prepares, packages, serves, vends or otherwise provides food for human consumption
- ▲ Develops or implements specific policies, procedures, or standards aimed at preventing foodborne illness
- ▲ Coordinates training, supervises or directs food preparation activities, and takes corrective action as needed to protect the health of the customer
- ▲ Conducts in-house, self-inspection of daily operations on a periodic basis to see that policies and procedures concerning food safety are being followed.

A certified food protection manager must demonstrate knowledge and skills in food protection management including the following areas:

- ▲ Identifying foodborne illness
- ▲ Describing the relationship between time and temperature and the growth of microorganisms that cause foodborne illness
- ▲ Describing the relationship between personal hygiene and food safety
- ▲ Describing methods for preventing food contamination from purchasing and receiving
- ▲ Recognizing problems and potential solutions associated with facility, equipment, and layout in a retail food establishment
- ▲ Recognizing problems and solutions associated with temperature control, preventing cross contamination, housekeeping and maintenance.

The Conference for Food Protection has adopted a complete list of knowledge areas presented in Appendix E.

Some state and municipal agencies already require certification of food managers, and others are considering it. Most certification programs require managers to pass a food safety examination recognized by the Conference for Food Protection. These examinations are designed to link food safety theory to practice. By passing a certification examination, the retail food manager is able to demonstrate proficiency in food protection management. Information about the Conference for Food Protection and its program for recognizing food protection manager certification examinations can be found at the organization's Web site at http://www.foodprotect.org

Some managers will voluntarily seek training before they take a certification examination. Some jurisdictions will require training before testing while others will require only the test. Training can involve participating in a formal course, or it could involve using a nontraditional approach such as home study, correspondence, distance learning, and interactive computer programming.

This book has been written as a teaching tool to be used in conjunction with traditional and nontraditional food manager certification programs. Like the manager certification examinations, it links theory to practice. The authors of this book recognize knowing about food safety is not enough. You must also be able to apply the principles and practices that enhance food safety during preparation, display and handling.

Summary

In the case study presented at the beginning of this chapter, you read about a foodborne disease outbreak caused by ground beef contaminated with Shiga toxin-producing *Escherichia coli* bacteria. These bacteria have been found in a variety of foods including raw ground beef. The bacteria are transferred to foods like ground beef through contact with the intestines of slaughtered animals. The risk of *E. coli* can be reduced by keeping meats at 41°F (5°C) or below during storage and display and by cleaning and sanitizing grinders and other equipment in the meat department on a regular basis. Retail food managers can also reduce the risk of Shiga toxin-producing *Escherichia coli* outbreaks by encouraging their customers to cook ground beef to an internal temperature of 160°F (71°C) for 15 seconds.

Everyone who works in a retail food establishment must understand the importance of food safety. It is the duty of every food establishment operator, manager, and employee to handle foods safely. Failure to do so can have serious financial impact on your establishment and may cost you your job.

You can protect the health and safety of your customers by developing and implementing effective food safety and sanitation practices within your establishment. In the following chapters, you will learn more about how food is contaminated and what actions are needed to prevent, control and eliminate the agents that frequently cause foodborne illness and spoilage.

Discussion Questions

1. Discuss the role of government in food safety programs.
2. Define the terms sanitary and contamination.
3. List some examples of food establishments. List some examples of retail food establishments.
4. How does news media coverage influence food safety issues?
5. What is a foodborne disease outbreak?
6. What impact do purchasing trends have on food safety?
7. How does food safety affect the young, elderly, pregnant women, and those with compromised immune systems?
8. Discuss how certification of food managers could affect food safety.
9. Identify some reasons to implement food safety programs in retail food establishments.
10. Discuss some reason(s) you are studying food safety and sanitation.

Quiz 1.1 (Multiple Choice)

1. Which of the following groups is not especially susceptible to foodborne illness?
   a. The very young.
   b. Young adults.
   c. The elderly.
   d. Pregnant or lactating women.
2. The cost of foodborne illness can occur in the form of:
   a. Medical expenses.
   b. Loss of sales.
   c. Legal fees and fines.
   d. All of the above.

3. Reports by the CDC show in most foodborne illness outbreaks mishandling of the suspect food occurred within which of the following stages?
   a. Transportation.
   b. Retail food establishments.
   c. Food manufacturing.
   d. Food production (farms, ranches, etc.).
4. If a piece of equipment is sanitary it:
   a. Is free of visible soil.
   b. Has been sterilized.
   c. Is a single-service item.
   d. Has had disease-causing germs reduced to safe levels.
5. How does the *FDA Food Code* affect individual states and jurisdictions?
   a. The Code is a federal law that must be enforced by state agencies.
   b. The Code regulates food manufacturing facilities (processors) in state jurisdictions.
   c. It provides a model for new laws and rules in state and local jurisdictions.
   d. It validates current practices.

Quiz 1.2 (True/False)

Answer questions 1-5 with either True or False.

1. The Hazard Analysis Critical Control Point System is only used to monitor food processing in manufacturing plants.
2. Certification of food protection managers and employees refers to screening done for health problems.
3. Regulatory agencies are assigned the task of monitoring the production and processing of food from harvest to the consumer.
4. The term retail food establishment includes any store selling a line of dry retail food, canned goods, non-food items and perishable foods.
5. Bacteria and viruses cause most foodborne illnesses.

Answers to the multiple choice and true/false questions are provided in Appendix A.

References/Suggested Readings

Centers for Disease Control and Prevention (2000). *Surveillance for Foodborne Disease Outbreaks—United States, 1993-1997.* U.S. Department of Health and Human Services. March 17, 2000. Atlanta, GA.

Food and Drug Administration (2001). *2001 Food Code.* U.S. Public Health Service. Washington, DC.

Jay, James M. (2000). *Modern Food Microbiology* (6th Edition). Aspen Publications. Gaithersburg, MD.

Mead, Paul S., Laurence Slutsker, Vance Dietz, Linda F. McCaig, Joseph S. Bresee, Craig Shapiro, Patricia M. Griffin, and Robert V. Tauxe (1999). "Food-Related Illness and Death in the United States." *Emerging Infectious Diseases,* Vol. 5, No. 5, September–October, 1999. Centers for Disease Control and Prevention. Atlanta, GA.

Potter, N., and J. Hotchkiss (1995) *Food Science* (3rd Edition). Aspen Publishers. Gaithersburg, MD.

Willard Bishop Consulting, 2001

Suggested Web Sites

Gateway to Government Food Safety Information
www.foodsafety.gov

United States Department of Agriculture (USDA)
www.usda.gov

Centers for Disease Control and Prevention (CDC)
www.cdc.gov

United States Food and Drug Administration (FDA)
www.fda.gov

USDA/FDA Food and Nutrition Information Center
www.nal.usda.gov/fnic/

Partnership for Food Safety Education
www.fightbac.org

The Food Marketing Institute
www.fmi.org

The Food Marketing Institute-Retail Food Establishment Industry Facts
www.fmi.org/facts_figs/superfact

Conference for Food Protection
www.foodprotect.org

2 Hazards to Food Safety

Foodborne Illness Linked to Foods Sold in Produce Department

Potentially hazardous foods, like meat, poultry, and seafood, top the list of food safety concerns. However, the number of cases of foodborne illness associated with fresh fruits and vegetables has increased significantly in the last 10 years. Produce is responsible for approximately 10% of the foodborne disease outbreaks each year. Shiga toxin-producing Escherichia coli *bacteria have been found in contaminated apple cider.* Salmonella *bacteria have contaminated cantaloupe. Other incidents include strawberries infected with Hepatitis A virus and raspberries tainted with* Cyclospora *parasite.*

Over the past decade, nearly 100 major food poisoning outbreaks were linked to contaminated sprouts, lettuce, and other produce. This is almost as many as were linked to beef. Why is there such an increased concern for fresh produce? What can retail food establishments do to reduce the food safety risk posed by the produce served and sold in their operations?

Learning Objectives

After reading this chapter, you should be able to:

- List the three main types of foodborne hazards.
- Identify the difference between infections, intoxications, and toxin-mediated infections as classes of foodborne illness.
- List the factors that promote the growth of disease-causing bacteria.
- Identify the food temperature danger zone.
- Explain how temperatures in the temperature danger zone influence bacterial growth.

▲ Identify the major types of potentially hazardous foods and the characteristics that are common to these foods.

Essential Terms

Acid	Pathogenic
Aerobic	pH
Alkaline	Physical hazards
Anaerobic	Potable water
Bacteria	Potentially hazardous foods
Biological hazards	Ready-to-eat (RTE) foods
Chemical hazards	Spoilage
Facultative anaerobic	Spore
FATTOM	Temperature abused
Food allergen	Toxin
Foodborne illness	Toxin-mediated infection
Hazard	Vegetative cells
Infection	Virus
Intoxication	Water activity (A_w)
Parasite	

Foodborne Illness

Have you ever had a foodborne illness? Many of us have, but it did not last long enough to be diagnosed. We think we have the "stomach flu" or a "24-hour bug" of some kind.

Did you become nauseated, feel abdominal pain and cramping, have diarrhea, and even start vomiting? It could be foodborne illness. When this happens, do you think it is from eating or drinking too much? If you are far from home, do you blame the "strange water"? Could it have been something you ate?

What you experience is often a foodborne illness (Figure 2.1). The symptoms are very similar to those associated with the flu. The type of microbe, how much contamination is in the food, and the general condition of the affected person determine the severity of the symptoms.

Figure 2.1 Ingredients for Foodborne Illness

General symptoms of foodborne illness usually include one or more of the following:

- Headache
- Nausea
- Vomiting
- Dehydration
- Abdominal pain
- Diarrhea
- Fatigue
- Fever.

Foodborne illness is generally classified as a foodborne infection, intoxication, or toxin-mediated infection (Figure 2.2). Your awareness of how different microbes cause foodborne illness will help you understand how they contaminate food.

When a living, disease-causing microorganism is eaten along with a food, it can cause a foodborne **infection**. After ingestion, the organism burrows into the lining of the victim's digestive tract and begins to grow in number. This can lead to the common symptoms of foodborne illness like diarrhea. Sometimes, the microbes may spread to other parts of the body through the blood stream. Bacteria, viruses, and parasites are examples of microorganisms that can cause infection. A common type of foodborne infection is salmonellosis. *Salmonella* bacteria frequently found in poultry and eggs cause this disease.

Infection	Caused by eating food that contains living, disease-causing microorganisms.
Intoxication	Caused by eating food that contains a harmful chemical or toxin produced by bacteria or other source.
Toxin-mediated Infection	Caused by eating a food that contains harmful microorganisms that produce a toxin once inside the human intestinal tract.

Figure 2.2 Classifications of Foodborne Illness

Intoxications are caused when a living organism multiplies in or on a food and produces a chemical waste or toxin. If the food containing the toxin is eaten, the toxin causes an illness. Foodborne intoxications are frequently called "food poisoning." Significant outbreaks have been caused by *Clostridium botulinum* and *Staphylococcus aureus* bacteria. An intoxication may also occur when an individual consumes food that contains man-made chemicals such as cleaning agents or pesticides.

A **toxin-mediated infection** is caused when a living organism is consumed with food (as in the case of an infection). Once the organism is inside the human body it produces a toxin that causes the illness. A toxin-mediated infection is different from intoxication because the toxin is produced inside the human body. An example of an organism that causes this type illness is *Clostridium perfringens*.

Foodborne illnesses have different onset times. The "**onset time**" is the time period between the time a person eats contaminated food and when they show the first symptoms of the disease. This is usually expressed in a range of hours since individual onsets vary depending on factors such as victim's age, health status, body weight, and the amount of contaminant they ingested with the food. The onset time for some foodborne diseases, such as Hepatitis A and trichinosis, can be several days.

Anyone can become ill from eating contaminated foods. In most cases, healthy adults will have flu-like symptoms and recover in a few days. The risks and dangers associated with foodborne illness are much more serious for people who are part of highly susceptible populations (see Figure 2.3). Highly susceptible populations include infants and young children, the elderly, pregnant women, individuals with suppressed immune systems due to acquired immune deficiency syndrome (AIDS), cancer, diabetes, and people taking certain types of medications. For these individuals, the symptoms and duration of foodborne illness can be much more severe–even life threatening.

Figure 2.3 Highly Susceptible Populations

Foodborne Hazards

A **foodborne hazard** refers to a biological, chemical, or physical hazard that can cause illness or injury when consumed along with the food (see Figure 2.4).

(Copyright Dennis Kunkel Microscopy, Inc.)

Figure 2.4 Foodborne Hazards

Biological hazards include bacteria, viruses, parasites, and fungi. These organisms are very small and can only be seen with the aid of a microscope. Biological hazards are commonly associated with humans and with raw products entering a retail food establishment. Bacteria are single-celled microorganisms that require food, moisture, and specific temperatures to multiply. **Bacteria** can cause foodborne infections, intoxications, and toxin-mediated infections. Many of these bacteria are present naturally in the environment where foods are grown. In retail food establishments, most are destroyed or controlled by monitoring time and temperature, good personal hygiene practices, an effective cleaning and sanitation program, and measures to minimize cross contamination.

Biological hazards **are by far the most important foodborne hazard in any type of retail food establishment.** They cause most foodborne illnesses and are the primary target of a food safety program.

Chemical hazards are toxic substances that may occur naturally or may be added during the processing of food. Examples of chemical contaminants include agricultural chemicals (i.e., pesticides, fertilizers, antibiotics), cleaning compounds, heavy metals (lead and mercury), food additives, and food allergens for allergen sensitive people. Harmful chemicals at very high levels have been associated with severe poisonings and allergic reactions. Chemicals and other non-food items should never be placed near food items.

Physical hazards are hard or soft foreign objects in food that can cause illness and injury. They include items such as fragments of glass, metal, unfrilled toothpicks, jewelry, adhesive bandages, and human hair. These hazards result from accidental contamination and poor food-handling practices that can occur at many points in the food chain from the source to the customer.

Bacteria

Bacteria and viruses are the most important biological foodborne hazards for any retail food establishment. Bacteria and viruses are responsible for more cases of foodborne illness than any other hazards.

All bacteria exist in a "vegetative state." **Vegetative cells** grow, reproduce, and produce wastes just like other living organisms.

Some bacteria have the ability to form structures called "spores." **Spores** help bacteria survive when their environment is too hot, cold, dry, acidic, or when there is not enough food. Spores are not able to grow or reproduce (see Figure 2.5).

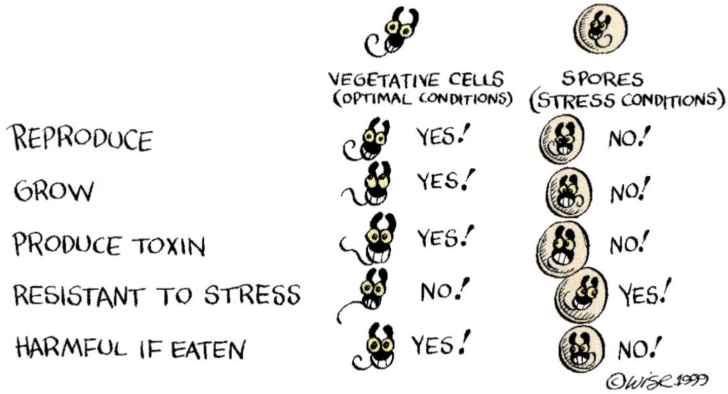

Figure 2.5 The Vegetative and Spore States of Bacteria Cells

However, when conditions become suitable for growth, a spore can germinate much like a seed. The bacterial spore can then return to the vegetative state and begin to grow again. Bacteria can survive for many months as spores. Also, it is much harder to destroy bacteria when they are in a spore form (Figure 2.6). The key to food safety is to destroy bacterial spores or keep them from changing into the dangerous vegetative state where they can grow.

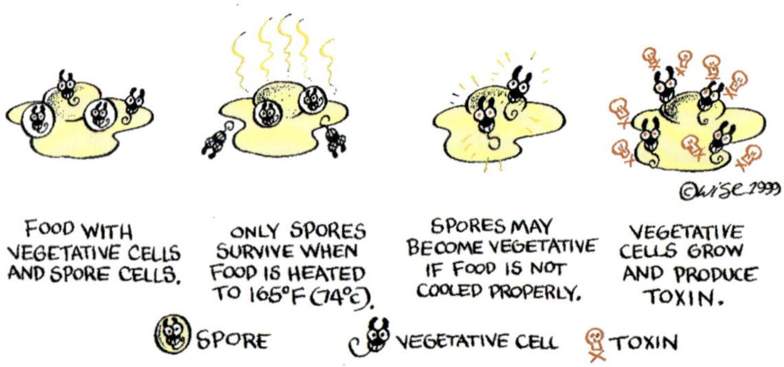

Figure 2.6 Spores and Toxins Pose a Special Challenge

Spoilage and Disease-causing Bacteria

Bacteria are classified as either spoilage or pathogenic (disease-causing) microorganisms.

Spoilage bacteria degrade (break down) foods so they look, taste, and smell bad. They reduce the quality of food to unacceptable levels. When this happens, the food may have to be thrown away creating a loss in profits.

Pathogenic bacteria are disease-causing microorganisms that can make people ill if the vegetative bacterial cells or their toxins are consumed with food.

Both spoilage and pathogenic bacteria must be controlled in retail food establishments. The steps you take to keep foods safe from disease-causing bacteria will also improve the culinary quality and shelf life of the food by controlling spoilage organisms.

Bacterial Growth

Bacteria reproduce when one bacterial cell divides to form two new cells. This process is called **binary fission**. The reproduction of bacteria and an increase in the number of organisms is referred to as bacterial growth (Figure 2.7).

Figure 2.7 Division of Bacterial Cells

Bacterial growth follows a regular pattern that consists of four phases (Figure 2.8). The first phase is the lag phase in which the bacteria exhibit little or no growth. The bacteria adjust to their surroundings during the lag phase. The lag phase lasts only a few hours at room temperature. However, the duration of this phase can be increased by keeping foods at 41°F (5°C) or below.

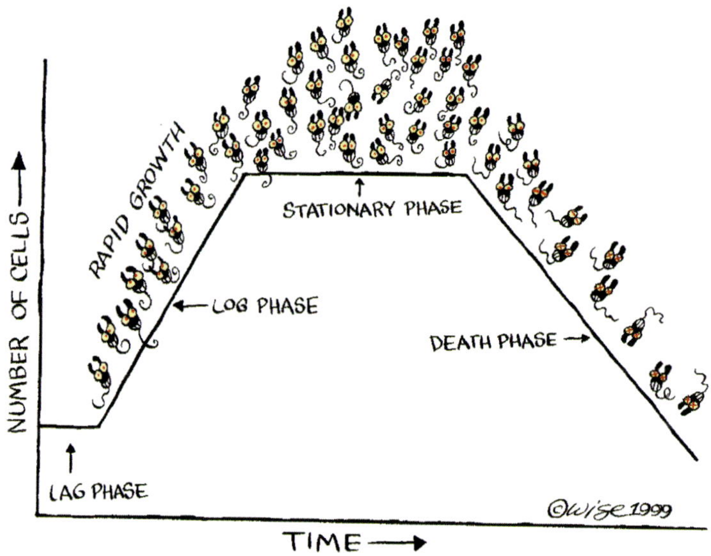

Figure 2.8 Bacterial Growth Curve

The second phase of bacterial growth is the log phase. Bacterial growth is very rapid during the log phase with bacteria doubling in numbers every few minutes. Keeping bacteria from reaching the log phase of growth is critical for food safety.

The third phase of bacterial growth is the stationary phase. The number of new bacteria being produced equals the number of organisms that are dying off during this phase. The bacteria have used up much of the space, nutrients, and moisture in the food by this phase of the growth curve.

The final phase in the growth curve is the death phase. Here bacteria die off rapidly because they lack nutrients and are poisoned by their own toxic wastes.

What Disease-causing Bacteria Need in Order to Multiply

Disease-causing bacteria need six conditions in order to multiply. They need a source of food, a mildly acidic environment (pH of 4.6 - 7.0), a temperature between 41°F (5°C) and 140°F (60°C), time of four hours or more, different oxygen-requiring environments, and enough moisture (water activity above .85). These requirements can be remembered using the acronym **F-A-T-T-O-M** (Figure 2.9). Since many foods naturally contain microorganisms, it is necessary to control one or more of these six conditions to prevent bacteria from multiplying.

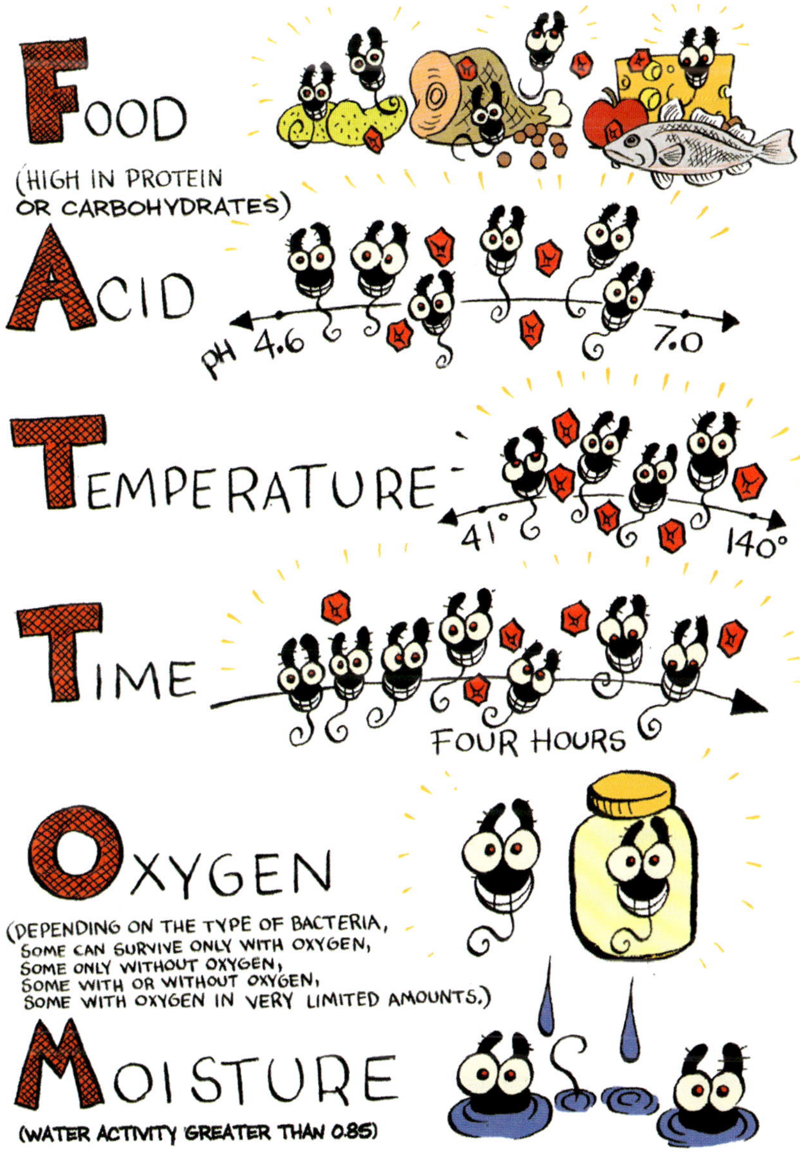

Figure 2.9 Six Conditions Bacteria Need to Multiply

Food

A suitable food supply is the most important condition needed for bacterial growth. Most bacteria prefer foods high in protein or carbohydrates like meats, poultry, seafood, dairy products, and cooked rice, beans, and potatoes (Figure 2.10).

Figure 2.10 Microbes Eat the Same Foods We Do

Acidity

The **pH** symbol is used to designate the level of acidity or alkalinity of a food. You measure pH on a scale that ranges from 0 to 14 (Figure 2.11).

Most foods are acidic and have a pH less than 7.0. Foods highly acidic (pH below 4.6), like lemons, limes, and tomatoes, will not normally support the growth of disease-causing bacteria. Pickling fruits and vegetables preserves the food by adding acids such as vinegar. This lowers the pH of the food in order to slow down the rate of bacterial growth.

A pH above 7.0 indicates the food is "**alkaline**." Only a few foods are alkaline. Examples of alkaline foods are olives, egg whites, and soda crackers.

Most bacteria that can cause foodborne illness prefer a neutral environment (pH of 7.0) but are capable of growing in foods that have a pH in the range of 4.6 to 9.0. Since most foods have a pH of less than 7.0, we have identified the range where harmful bacteria grow as 4.6 to 7.0. Many foods offered for sale in retail food establishments have a pH in this range.

Disease-causing bacteria grow best when the foods they live in and on have a pH of 4.6 to 7.0. Milk, meat, poultry, fish, and dairy products are in this range.

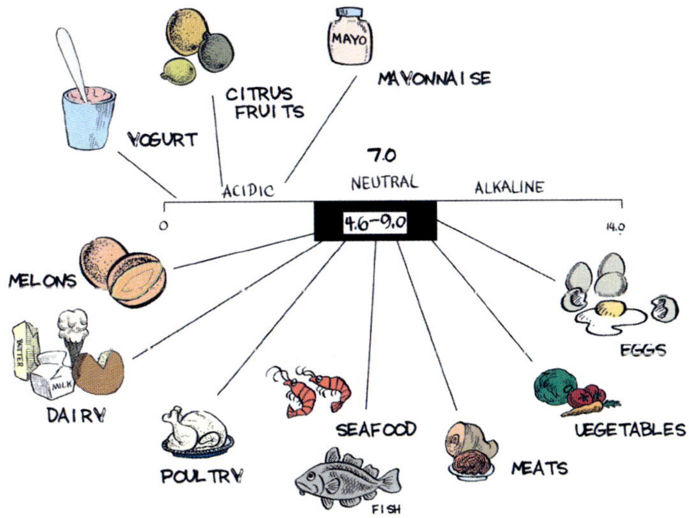

Figure 2.11 Acidity Range of Common Foods Prepared in Food Establishments

Temperature

Temperature in retail food establishments is commonly measured using a food thermometer scaled in degrees Fahrenheit (denoted as °F), degrees Celsius (denoted as °C), or both. Throughout this book, temperatures will be provided in both Fahrenheit and Celsius to acquaint you with both scales. For example, the temperature requirement for cold-holding or cold display of potentially hazardous foods will be shown in both scales as 41°F (5°C).

All bacteria do not have the same temperature requirements for growth. **Psychrophilic** bacteria grow within a temperature range of 32°F (0°C) to 70°F (21°C). These microorganisms are especially troublesome because they are capable of multiplying at both refrigerated and room temperatures. Most bacteria that grow well at refrigerated temperatures are spoilage organisms, but some, like *Listeria monocytogenes,* can cause disease.

Mesophilic (middle range) bacteria grow at temperatures between 70°F (21°C) and 110°F (43°C), with most rapid growth at human body temperature of 98.6°F (37°C).

Bacteria that grow best at temperatures above 110°F (43°C) are called **thermophilic** organisms. Thermophilic bacteria are spoilage organisms.

Time and temperature are the most critical factors affecting the growth of bacteria in foods. Most disease-causing bacteria can grow within a temperature range of 41°F (5°C) to 140°F (60°C). This is commonly referred to as the food "**Temperature Danger Zone**" (Figure 2.12). A few disease-causing bacteria, such as *Listeria monocytogenes,* can grow at temperatures below 41°F (5°C), but the rate of growth is very slow.

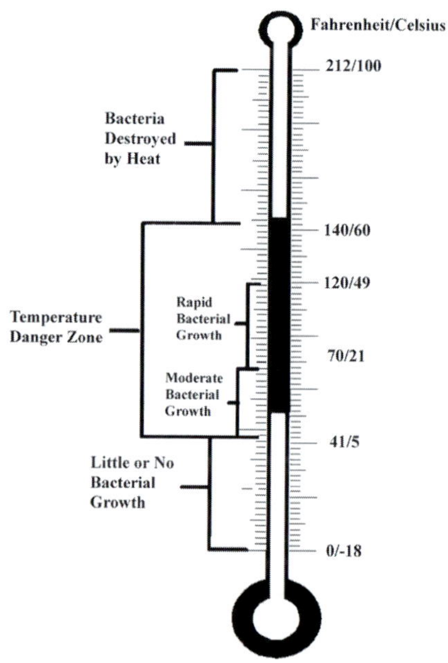

Figure 2.12 Temperature Control Guide

Careful monitoring of time and temperature is the most effective way a retail food establishment has to control the growth of disease-causing and spoilage bacteria. A saying related to temperature control in the retail food industry is "Keep it hot, keep it cold, or don't keep it!" This means all cold foods must be stored at 41°F (5°C) or below and all hot foods held at 140°F (60°C) or above. Temperature abuse is the term applied to foods that have not been heated to a safe temperature or kept at the proper temperature to control growth. These mistakes could result in a foodborne illness.

Time

Under ideal conditions, bacterial cells can double in number every 15 to 30 minutes. *Clostridium perfringens* bacteria can double every 10 minutes (Longreè and Armbruster, 1996). For most bacteria, a single cell can generate over one million new cells in just five hours (Figure 2.13). It is very important not to give bacteria an opportunity to multiply. Control of temperature is critical in preventing bacterial growth.

Because bacteria have the ability to multiply rapidly, it does not take long before many cells are produced. A rule of thumb in the retail food industry is **bacteria need about four hours to grow to high enough numbers to cause illness**. This includes the total time a food is between 41° (5°C) and 140°F (60°C). Remember, a single bacterial cell can produce over one million new cells in just five hours under ideal conditions.

Time	0	15 min.	30 min.	60 min.	3 hrs.	5 hrs.
# of cells	1	2	4	16	>1000	>1 million

Figure 2.13 Bacterial Growth Time

Oxygen

Bacteria also differ in their requirements for oxygen. **Aerobic** bacteria must have oxygen in order to grow. These microbes normally grow on the outside of food where oxygen is available. A good example of aerobic bacteria would be *Pseudomonas spp.* that causes spoilage and produces a slimy feeling on the outside of deli lunch meats.

Anaerobic bacteria, however, cannot survive when oxygen is present because it is toxic to them. These bacteria grow well in vacuum packaged foods or canned foods where oxygen is not available. Anaerobic conditions also exist in the middle of cooked food masses such as in large stockpots, baked potatoes, or in the middle of a roast or ham.

Facultative anaerobic forms of bacteria can grow with or without free oxygen but have a preference. Most foodborne disease-causing microorganisms are facultative anaerobes. Controlling oxygen conditions is not an effective way to prevent foodborne illness. Regardless of available oxygen, some disease-causing bacteria will be able to adapt to the conditions and grow.

Moisture

Like most other forms of life, moisture is an important factor in bacterial growth. That is why man has dried foods for thousands of years as a way to preserve them.

Scientists have determined it is not the percentage of moisture or "water by volume" in a food that most affects bacterial growth. Rather it is the amount of "available water" for bacterial activity. The amount of available water in a food is expressed as water activity and is designated with the symbol A_w. **Water activity** is a measure of the amount of water that is not bound to the food and is therefore available to support bacterial growth (Figure 2.14).

For example, fresh chicken has 60% water by volume, and its A_w is approximately .98. The same chicken, when frozen, still has 60% water by volume but its A_w is nearly zero. Water activity is measured on a scale from 0 to 1.0. **Disease-causing bacteria can only grow in foods that have a water activity higher than .85**. Lowering their water activity to .85 or below preserves many foods. Drying foods or adding salt or sugar reduces the amount of available water. For example, jams and jellies that contain a lot of sugar have an A_w much less than .85. This alone prevents the growth of disease-causing microorganisms.

Potentially Hazardous Foods (PHF)

Potentially hazardous foods are food items that require temperature control because they are capable of supporting the rapid and progressive growth of infectious or toxin-producing microbes, the growth and toxin production of *Clostridium botulinum* bacteria, or the growth of *Salmonella enteritidis* bacteria in raw shell eggs.

Figure 2.14 Water Activity (A_w) of Some Foods Sold in Food Establishments

The *FDA Food Code* classifies the following natural and man-made items as potentially hazardous foods:

- ▲ Foods of animal origin that are raw or heat-treated
- ▲ Foods of plant origin that are heat-treated or consist of raw seed sprouts
- ▲ Cut melons
- ▲ Garlic-in-oil mixtures that are not modified in a way to inhibit the growth of disease-causing microorganisms.

Common examples of potentially hazardous foods are red meats, poultry, shell eggs that have not been treated to destroy all viable *Salmonella* bacteria, fish and shellfish, and dairy products (Figure 2.15). Other examples of potentially hazardous foods include vegetables such as cooked rice and potatoes, and fruits such as cut cantaloupe.

If potentially hazardous foods are held in the temperature danger zone [between 41°F (5°C) and 140°F (60°C)] for four hours or more, infectious and toxin-producing microbes can grow to dangerous levels. Potentially hazardous foods have been associated with most foodborne disease outbreaks. It is critical to control the handling and storage of potentially hazardous foods to prevent bacterial growth.

Figure 2.15 Examples of Potentially Hazardous Foods

Ready-to-Eat Foods

Ready-to-eat foods are food items that are edible without washing, cooking or additional preparation by the customer or by the retail food establishment. These foods can become contaminated if not handled properly. The *FDA Food Code* identifies the following types of food as ready-to-eat:

- ▲ Raw animal foods that are cooked (i.e., rotisserie chicken) or frozen (i.e., sushi) as prescribed in the *FDA Food Code*
- ▲ Raw fruits and vegetables that are washed
- ▲ Fruits and vegetables that are cooked for hot-holding
- ▲ All potentially hazardous foods that are cooked and then cooled as prescribed in the *FDA Food Code*
- ▲ Plant foods for which further washing, cooking, or other processing is not required for food safety, and from which rinds, peels, husks, or shells, if naturally present, are removed
- ▲ Substances derived from plants such as spices, seasonings, and sugar

- ▲ Bakery items such as bread, cakes, pies, fillings, or icing for which further cooking is not required for food safety
- ▲ Dry, fermented sausages (i.e., dry salami or pepperoni), salt-cured meat and poultry products (i.e., prosciutto ham, country cured ham, and Parma ham), and dried meat and poultry products (i.e., jerky or beef sticks) that are produced in accordance with USDA guidelines and have been treated to destroy pathogens
- ▲ Thermally processed low acid foods (i.e., smoked fish or meat) packaged in hermetically sealed containers.

Foodborne Illness Caused by Bacteria

Biological hazards are much more important for the retail food establishment manager to control because they lead to the majority of foodborne illness. Biological hazards include bacteria, viruses, parasites, and fungi and are the most common agents that lead to foodborne illness.

Bacteria are classified as sporeforming and non-sporeforming organisms. In the following sections, each type of biological hazard is described, the common foods and route of transmission are identified, and preventive strategies are discussed. In Figure 2.16, common biological hazards are listed.

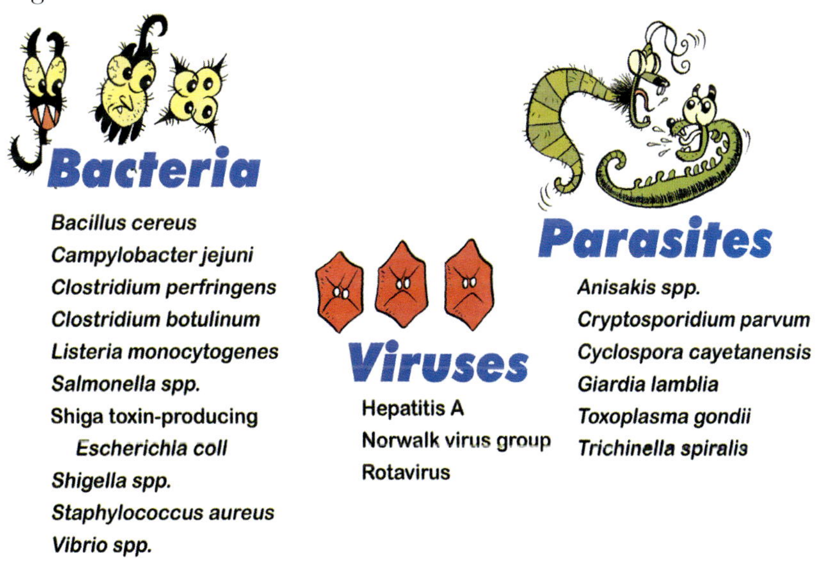

Figure 2.16 Common Biological Hazards in Food Establishments

Foodborne Illness Caused by Sporeforming Bacteria

The following group of bacteria is capable of forming spores. Remember, a spore structure enables a cell to survive environmental stress such as cooking, freezing, high salt conditions, drying, and high acid conditions (Figure 2.17).

Spores are not harmful if ingested, except in a baby's digestive system, where *Clostridium botulinum* spores can cause a disease called infant botulism. It is often recommended parents avoid serving honey to babies due to the possible presence of *Clostridium botulinum* spores. If conditions in the food are suitable for bacterial growth and the spore turns into a vegetative cell, the vegetative cell can grow in the food and cause illness if eaten.

Figure 2.17 Well-Protected Spores

Sporeforming bacteria are generally found in foods that are grown in soil, like vegetables and spices. They may also be found in animal products. They can be particularly troublesome in retail food establishments when foods are not cooled properly. Even proper cooking or reheating may not destroy spores.

For example, a 10-gallon pot of chili was prepared for the next day's salad bar display. All the ingredients (beans, meat, spices, tomato base) were mixed together and cooked to a rapid boil. Vegetative cells should die, but spores may survive. Refer to Figure 2.6 on page 32 to see how this happens.

The chili was then stored in the 10 gallon pot and allowed to cool overnight in a walk-in refrigerator. It can take the core temperature of the chili two to three days to cool from 140°F (60°C) to 41°F (5°C)! If given enough time at the right temperature during the cooling process, sporeforming bacteria that survived the cooking process may change into vegetative cells and begin to grow.

Spores are most likely to turn into the dangerous vegetative state when:

- ▲ They are "heat-shocked" during cooking which can allow the spores to become vegetative cells
- ▲ Optimum conditions exist for growth (high protein or carbohydrates, high moisture, pH greater than 4.6)
- ▲ Temperatures are in the food temperature danger zone or between 41°F (5°C) to 140°F (60°C) for four or more hours.

To prevent sporeforming bacteria from changing to the dangerous vegetative state, it is critical hot-food temperatures be maintained at 140°F (60°C) or above and cold foods should be held at 41°F (5°C) or below. Always cook and cool foods as rapidly as possible (within four hours) to limit bacterial growth.

Bacillus cereus

CAUSATIVE AGENT (*sporeforming bacteria)	TYPE OF ILLNESS	SYMPTOMS ONSET	COMMON FOODS	PREVENTION
**Bacillus cereus*	Bacterial intoxication or toxin-mediated infection	1) Diarrhea type: abdominal cramps (8 to 16 hrs) 2) Vomiting type: vomiting, diarrhea, abdominal cramps (30 min to 6 hrs)	1) Diarrhea type: meats milk vegetables 2) Vomiting type: rice starchy foods grains cereals	Properly heat, cool, and reheat foods; proper cooling is most important

Description—*Bacillus cereus* is a sporeforming bacterium that can survive with or without oxygen. It has been associated with two very different types of illnesses: one vomiting, the other diarrhea.

Transmission in Foods—Illness due to *Bacillus cereus* is most often attributed to foods that are improperly stored (cooled, hot-held), permitting the conversion of spores to vegetative cells. Vegetative cells then produce toxin in the food that leads to illness.

Prevention—Foods, such as rice, must be cooked and held hot at 140°F (60°C) or above if not consumed immediately. These foods must also be cooled rapidly to below 41°F (5°C) prior to storage. Failure to keep foods out of the temperature danger zone provides an environment that promotes the growth of *Bacillus cereus* bacteria. As the bacteria grow, they produce toxins that can cause illness.

Clostridium perfringens

CAUSATIVE AGENT (*sporeforming bacteria)	TYPE OF ILLNESS	SYMPTOMS ONSET	COMMON FOODS	PREVENTION
**Clostridium perfringens*	Bacterial toxin-mediated infection	Intense abdominal pains and severe diarrhea, (8 to 22 hrs)	Spices, gravy, improperly cooled foods (especially meats and gravy dishes)	Properly cook, cool, and reheat foods; proper cooling is most important

Description—*Clostridium perfringens* is a nearly anaerobic (must have very little oxygen), sporeforming bacterium that causes foodborne illness. Potentially hazardous foods that have been temperature abused [not kept hot – above 140°F (60° C); or cold – below 41°F (5°C)] are frequently associated with this problem. *Clostridium perfringens* causes illness due to a toxin-mediated infection where the ingested cells colonize and then produce a toxin in the human intestinal tract.

Transmission in Foods—Illness due to *Clostridium perfringens* is most often attributed to foods that are temperature abused, especially those that have been improperly cooled and reheated.

Prevention—Foods must be cooked to 145°F (63°C) or above. Cooked foods must be cooled from 140°F (60°C) to 70°F (21°C) within two hours and from 70°F (21°C) to 41°F (5°C) in an additional four hours. Foods must also be reheated to 165°F (74°C) within two hours and held at 140°F (60°C) until served. For quality and safety reasons, foods should be reheated only once.

Clostridium botulinum

CAUSATIVE AGENT (*sporeforming bacteria)	TYPE OF ILLNESS	SYMPTOMS ONSET	COMMON FOODS	PREVENTION
**Clostridium botulinum*	Bacterial intoxication	Dizziness, double vision, difficulty in breathing and swallowing, headache (12 to 36 hrs)	Improperly canned foods, vacuum packed refrigerated foods, cooked foods in anaerobic mass	Properly heat-process anaerobically packed foods; DO NOT use home canned foods; DO NOT vacuum package potentially hazardous foods; be sure to follow safe storage and display temperature guidelines for reduced oxygen foods

Description—*Clostridium botulinum* is an anaerobic (must not have oxygen), sporeforming bacterium that causes foodborne intoxication due to improperly heat-processed foods (especially home canning). The organism produces a neurotoxin that is one of the deadliest biological toxins known to man. This toxin is not heat stable and can be destroyed if the food is boiled for about 20 minutes. However, botulism still occurs because people do not want to boil food that has already been cooked.

Transmission in Foods—Illness due to *Clostridium botulinum* is almost always attributed to ingestion of foods that were not heat processed correctly and packaged anaerobically.

Prevention—Do not use home-canned foods or buy foods from unapproved sources. Do not can foods in a retail food establishment. The risk is too great. If another form of reduced oxygen packaging is used, such as vacuum packaging of smoked ham, use extreme caution. Discuss the potential risks with a food safety specialist and/or your regulatory authority. If the proper controls are not in place, this type of packaging can allow growth of *Clostridium botulinum* cells. Follow temperature guidelines for vacuum packaged and other reduced oxygen packaged foods. Discard damaged cans (see Chapter 4). The *FDA Food Code* stipulates a variance is required for any retail food establishment that prepares products in reduced-oxygen packaging. Canned and bottled foods are not the only place you can find *Clostridium botulinum* bacteria. The anaerobic conditions created in hot-food masses, which are not cooled properly, also provide an environment that is suitable for these organisms.

Foodborne Illness Caused by Non-sporeforming Bacteria

The following group of bacteria does not form spores. Compared to bacterial spores, vegetative cells are easily destroyed by proper cooking. There are numerous examples of non-sporeforming foodborne bacteria that are important in the food industry.

Campylobacter jejuni

CAUSATIVE AGENT	TYPE OF ILLNESS	SYMPTOMS ONSET	COMMON FOODS	PREVENTION
Campylobacter jejuni	Bacterial infection	Watery, bloody diarrhea (2 to 5 days)	Raw chicken, raw milk, raw meat	Properly handle and cook foods; avoid cross contamination

Description—*Campylobacter jejuni* is a major cause of foodborne infection. Many scientists and regulatory officials believe *Campylobacter jejuni* is the most frequent cause of foodborne illness. This organism requires a very strict amount of air for growth. It can tolerate only 3 to 6% oxygen for growth.

Transmission in Foods—*Campylobacter jejuni* is often transferred from raw meats to other foods by cross contamination. This typically involves the transfer of germs by a contaminated food-contact surface (such as a cutting board or knife) or a food employee's hands.

Prevention—Cook raw meats and poultry properly. For example, rotisserie chicken must be cooked to 165°F (74°C) for 15 seconds to destroy these organisms. Other preventative measures include cleaning and sanitizing food-contact surfaces (cutting boards) when changing from raw to cooked or ready-to-eat foods and proper hand washing after handling raw foods.

Shiga toxin-producing *Escherichia coli*

CAUSATIVE AGENT	TYPE OF ILLNESS	SYMPTOMS ONSET	COMMON FOODS	PREVENTION
Shiga toxin-producing *Escherichia coli*	Bacterial infection or toxin-mediated infection	Bloody diarrhea followed by kidney failure and hemolytic uremic syndrome (HUS) in severe cases (12 to 72 hrs)	Undercooked hamburger, raw milk, unpasteurized apple cider, fruit drinks, lettuce	Practice good food sanitation, handwashing, properly handle and cook foods

Description—The Shiga toxin-producing *Escherichia coli* group of bacteria includes any *E. coli* capable of producing shiga toxins. This includes but is not limited to *E. coli* reported as serotype O157:H7, O157:NM, and O157:H-. This facultative anaerobic bacteria can be found in the intestines of warm-blooded animals, especially cows. The illness caused by Shiga toxin-producing *Escherichia coli* can be an infection or a toxin-mediated infection. Only a small amount of bacteria is required to produce an illness. So, a potentially hazardous food is not needed to allow for bacterial growth. A low level of contamination is enough. That's why apple cider, which has a pH lower than 4.6, has been implicated in cases of foodborne illness.

Transmission in Foods—Shiga toxin-producing *Escherichia coli* is usually transferred to foods like beef through contact with the intestines of slaughtered animals. Apples used for juice from orchards where cattle grazed are also suspected. Transmission can occur if employees who are carriers do not wash their soiled hands properly after going to the toilet. Cross contamination by soiled equipment and utensils may also spread Shiga toxin-producing *Escherichia coli.*

Prevention—Cook ground meats to an internal temperature of at least 155°F (68°C) for 15 seconds. Make sure employees wash their hands thoroughly before starting to work with food and after going to the toilet. Prevent cross contamination and keep hot foods above 140°F (60°C) and cold foods below 41°F (5°C). Wash lettuce in sinks used only for food preparation. Use only pasteurized apple cider or fruit juice and milk products.

Listeria monocytogenes

CAUSATIVE AGENT	TYPE OF ILLNESS	SYMPTOMS ONSET	COMMON FOODS	PREVENTION
Listeria monocytogenes	Bacterial infection	1) Healthy adult: flu-like symptoms 2) At-risk population: septicemia, meningitis, encephalitis, birth defects 3) Still-birth (1 day to 3 weeks)	Raw milk, dairy items, raw meats, refrigerated ready-to-eat foods, processed ready-to-eat meats (hot dogs, luncheon meat), raw vegetables, and seafood	Properly store and cook foods; avoid cross contamination; rotate processed refrigerated foods using First In, First Out (FIFO) to ensure timely use; clean and sanitize refrigerated storage and display units regularly

Description—*Listeria monocytogenes* is a facultative anaerobic (can grow with or without oxygen) bacterium that causes foodborne infection. This microbe is important to retail food establishment operations because it has the ability to survive under many conditions such as in high-salt foods and, unlike most other foodborne pathogens, can grow at refrigerated temperatures below 41°F (5°C).

Transmission in Foods—Transmission to foods can occur by cross contamination by people or equipment or if foods are not cooked properly.

Prevention—Cook foods thoroughly and practice good food-handling techniques. Specific practices for control are timely use and rotation of refrigerated products, such as cooked turkey breasts, hot dogs, hams, and luncheon meats. Check refrigerated storage and display units for out-of-date items. Clean and sanitize units regularly to prevent possible contamination.

Salmonella spp.

CAUSATIVE AGENT	TYPE OF ILLNESS	SYMPTOMS ONSET	COMMON FOODS	PREVENTION
Salmonella spp.	Bacterial infection	Nausea, fever, vomiting, abdominal cramps, diarrhea (6 to 48 hrs)	Raw meats, raw poultry, eggs, milk, dairy products	Properly cook foods; avoid cross contamination

Description—*Salmonella* are facultative anaerobic (grow with or without oxygen) bacteria frequently implicated as a foodborne infection. These organisms are commonly found in the intestinal tract of humans and warm-blooded animals.

Transmission in Foods—Transmission in foods is commonly through cross contamination where fecal material is transferred to food through contact with raw foods (especially poultry), contaminated food-contact surfaces (i.e., cutting boards), or infected food employees. The incidence of salmonellosis has risen more than twenty-fold since 1946. One reason may be the increased number of centrally processed foods along with bulk distribution (Longreè and Armbruster, 1996).

Prevention—Cook foods thoroughly. For example, *Salmonella* bacteria will be destroyed when poultry is cooked to an internal temperature of 165°F (74°C) for 15 seconds and when eggs are cooked to 145°F (63°C) for 15 seconds. Clean and sanitize food-contact surfaces after use with raw animal foods, and make sure food employees wash their hands adequately before working with food, especially after going to the toilet or handling raw meats and poultry.

Shigella spp.

CAUSATIVE AGENT	TYPE OF ILLNESS	SYMPTOMS ONSET	COMMON FOODS	PREVENTION
Shigella spp.	Bacterial infection	Bacillary dysentery, diarrhea, fever, abdominal cramps, dehydration (1 to 7 days)	Foods that are prepared with human contact: salads, raw vegetables, milk, dairy products, raw poultry, non-potable water, ready-to-eat meat	Wash hands and practice good personal hygiene; properly cook foods

Description—*Shigella* are facultative anaerobic bacteria that account for about 10% of foodborne illnesses in the United States. These organisms are commonly found in the intestines and feces of humans and warm blooded animals. They cause shigellosis, a foodborne infection. The bacterium produces a toxin that causes watery diarrhea.

Transmission in Foods—Water that is contaminated by fecal material and food and utensils handled by employees who are carriers of the bacteria can cause this problem. Illness from *Shigella* is most often attributed to contaminated ready-to-eat foods handled by an infected food handler.

Prevention—Do not allow individuals who have been diagnosed with shigellosis to handle food. Make certain food employees wash their hands thoroughly after going to the toilet. Cook foods to proper temperature and do not allow cross contamination. Wash produce and other foods with potable water (water that is safe to drink).

Staphylococcus aureus

CAUSATIVE AGENT	TYPE OF ILLNESS	SYMPTOMS ONSET	COMMON FOODS	PREVENTION
Staphylococcus aureus	Bacterial intoxication	Nausea, vomiting, abdominal cramps, headaches (2 to 6 hrs)	Foods that are prepared with human contact, cooked or processed foods	Wash hands and practice good personal hygiene; cooking WILL NOT inactivate the toxin

Description—*Staphylococcus aureus* is a facultative anaerobic bacterium that produces a heat-stable toxin as it grows on foods. This bacterium can also grow on cooked and otherwise safe foods that are recontaminated by food employees who mishandle the food. *Staphylococcus aureus* bacteria do not compete well when other types of microorganisms are present. However, they grow well when alone and without competition from other microbes. *Staphylococcus aureus* bacteria produce toxins that cause an intoxication. These bacteria are commonly found on human skin, hands, hair, and in the nose and throat. They may also be found in burns, infected cuts and wounds, pimples, and boils. At any given time, *Staphylococcus aureus* bacteria can be found living in and on 30 to 50% of people, including those who are otherwise healthy. These organisms can be transferred to foods easily, and they can grow in foods that contain high salt or high sugar and a lower water activity, such as hams and luncheon meats.

Transmission in Foods—Since humans are the primary reservoir, contamination from the employee's hands is the most common way the organism is introduced into foods. Foods requiring considerable food preparation and handling are especially susceptible. The bacteria are also spread by droplets of saliva from talking, coughing, and sneezing near food. Food employees who improperly use tasting spoons and ladles can transfer bacteria from their mouth to food. Contaminated human hands combined with temperature abuse usually cause most problems associated with *Staphylococcus aureus.*

Prevention—Avoid contamination of food from bare hands. Do not allow individuals who have infected cuts, burns, or wounds to handle food unless the wound has been properly cleaned, bandaged and a plastic glove or similar barrier is used to cover the bandaged area. Make certain food employees wash their hands thoroughly before starting to work and whenever their hands become contaminated. Do not allow food employees to reuse tasting spoons and ladles. Heat and cool foods properly and keep ready-to-eat foods out of the temperature danger zone.

Vibrio spp.

CAUSATIVE AGENT	TYPE OF ILLNESS	SYMPTOMS ONSET	COMMON FOODS	PREVENTION
Vibrio spp.	Bacterial infection	Headache, fever, chills, diarrhea, vomiting, severe electrolyte loss, gastroenteritis (2 to 48 hrs)	Raw or improperly cooked fish and shellfish	Practice good sanitation; properly cook foods; implement procedures to separate raw and ready-to-eat seafood in display cases

Description—There are three organisms within the *Vibrio* group of bacteria that have been connected with foodborne infections. They include *Vibrio cholera*, *Vibrio parahaemolyticus*, and *Vibrio vulnificus*. All are important since they are very resistant to salt and are common in seafood.

Transmission in Foods—Since the organism is inherent in many raw seafood, transmission to other foods by cross contamination is a concern. Most illnesses are caused by the consumption of raw or undercooked seafood.

Prevention—Buy seafood from approved sources only. Be sure display cases do not allow for cross contamination from raw to ready-to-eat seafood. Cook seafood to the proper temperature and avoid eating raw or lightly cooked seafood, especially if you are in the at-risk segment of the population. Food employees must practice good personal hygiene and use correct handwashing procedures especially when handling raw seafood, then ready-to-eat foods.

Foodborne Illness Caused by Viruses

The viruses that cause foodborne disease differ from foodborne bacteria in several ways. **Viruses** are much smaller than bacteria, and they require a living host (human, animal) to reproduce. Viruses do not multiply in foods. However, a susceptible person needs to consume only a few viral particles in order to experience an infection.

Viruses are usually transferred from one food to another, from a food employee to a food, or from a contaminated water supply to a food. **A potentially hazardous food is not needed to support survival of viruses**.

There are three viruses of primary importance to retail food establishments: Hepatitis A, Norwalk virus, and Rotavirus. **Proper handwashing and separation of raw and ready-to-eat foods are important keys to controlling the spread of foodborne viruses.**

Hepatitis A

CAUSATIVE AGENT	TYPE OF ILLNESS	SYMPTOMS ONSET	COMMON FOODS	PREVENTION
Hepatitis A	Viral infection	Fever, nausea, vomiting, abdominal pain, fatigue, swelling of the liver, jaundice (10 to 50 days)	Foods that are prepared with human contact, contaminated water	Wash hands and practice good personal hygiene; separate raw and ready-to-eat seafood products in refrigerated display cases

Description—Hepatitis A is a foodborne virus that has been associated with many foodborne infections. Hepatitis A causes a liver disease called infectious hepatitis. The hepatitis virus is a particularly important hazard to retail food establishments because employees can harbor the virus for up to six weeks and not show symptoms of illness. Food employees are contagious for one week before onset of symptoms and two weeks after the symptoms of the disease appear. During that time, infected employees can contaminate foods and other employees by spreading fecal material from unwashed hands and nails. Hepatitis A virus is very hardy and can live for several hours in a suitable environment.

Transmission in Foods—The virus is transmitted by ingestion of food and water that contain the Hepatitis A virus. Raw seafood and foods handled by infected human hands are the largest threat of transmission and disease from Hepatitis A.

Prevention—Buy clams, oysters, and molluscan shellfish from approved sources. Raw and ready-to-eat foods must be kept separate during storage and display. Handle foods properly and cook them to recommended temperatures. Additional information about control and prevention of Hepatitis A is available in the *Hepatitis A Information Guide* published by the Food Marketing Institute and listed in the references/ suggested readings information at the end of this chapter.

Norwalk virus

CAUSATIVE AGENT	TYPE OF ILLNESS	SYMPTOMS ONSET	COMMON FOODS	PREVENTION
Norwalk virus	Viral infection	Vomiting, diarrhea, abdominal pain, headache, low grade fever, onset (24 to 48 hrs)	Sewage, contaminated water, contaminated salad ingredients, raw clams, oysters, and infected food employees	Wash hands and practice good personal hygiene; separate raw and ready-to-eat seafood products in refrigerated display cases

Description—The Norwalk virus is another common foodborne virus that has been associated with many foodborne infections.

Transmission in Foods—The virus is primarily transmitted by ingestion of food and water that has been contaminated with feces that contain the Norwalk virus.

Prevention—Use potable water (safe to drink) when preparing shellfish. Handle foods properly and cook foods to the correct temperature. Food employees must practice good personal hygiene and wash hands and fingernails thoroughly before working with food and after going to the toilet. Keep raw and ready-to-eat seafood products separate.

Rotavirus

CAUSATIVE AGENT	TYPE OF ILLNESS	SYMPTOMS ONSET	COMMON FOODS	PREVENTION
Rotavirus	Viral infection	Diarrhea (especially in infants and children), vomiting, low grade fever, 1 to 3 days onset, lasts 4 to 8 days	Sewage, contaminated water, contaminated salad ingredients, raw seafood	Good personal hygiene and handwashing; proper food handling practices

Description—Group A rotaviruses cause several diseases known as rotavirus gastroenteritis. Group A rotavirus is the leading cause of severe diarrhea among infants and children. Over three million cases of rotavirus gastroenteritis occur annually in the United States.

Transmission in Foods—Contaminated hands may spread the virus to foods that will not be cooked before eating.

Prevention—Cook foods to the proper temperature and handle properly. Food employees must practice good personal hygiene and wash hands and fingernails thoroughly before working with food and after going to the toilet.

Foodborne Illness Caused by Parasites

Foodborne parasites are another important foodborne biological hazard. **Parasites** are small or microscopic creatures that need to live on or inside a living host to survive. There are many parasites that can enter the food system and cause foodborne illness. In this chapter, we list a few of the most troublesome ones that may appear in retail food establishments. Parasitic infection is far less common than bacterial or viral foodborne illnesses.

Anisakis spp.

CAUSATIVE AGENT	TYPE OF ILLNESS	SYMPTOMS ONSET	COMMON FOODS	PREVENTION
Anisakis spp.	Parasitic infection	Coughing, vomiting, (1 hour to 2 weeks)	Raw or undercooked seafood, especially bottom feeding fish	Cook fish to the proper temperature throughout; freeze to meet *FDA Food Code* specifications

Description—*Anisakis spp.* are nematodes (roundworms) associated with foodborne infection from fish. The worms are about 1½ inches long and the diameter of a human hair. They are beige, ivory, white, gray, brown, or pink. Other names for this parasite are "cod worm" (not to be confused with common roundworms found in cod) and "herring worm."

Transmission in Foods—The natural hosts of the parasite are walruses, and perhaps sea lions and otters. The worms are transferred to fish, their intermediate host, in the water in which the walruses live. Humans become the accidental host upon eating fish that are infested with the parasites. Humans do not make good hosts for the parasites. The worms will not complete their life cycles in humans and eventually die.

Prevention—Inspect seafood and handle it carefully to expose worms. Purchase seafood from an approved supplier. Cook to proper temperatures or freeze to meet *FDA Food Code* specifications.

Cyclospora cayentanensis

CAUSATIVE AGENT	TYPE OF ILLNESS	SYMPTOMS ONSET	COMMON FOODS	PREVENTION
Cyclospora cayentanensis	Parasitic infection	Watery and explosive diarrhea, loss of appetite, bloating (1 week)	Water, strawberries, raspberries, raw vegetables	Good sanitation; reputable supplier

Description—*Cyclospora cayentanensis* is a parasite that has been reported much more frequently beginning in the 1990s. The most recent outbreaks of cyclosporiasis have been associated with fresh fruits and vegetables that were contaminated at the farm.

Transmission in Foods—*Cyclospora* is passed from person to person by fecal-oral transmission. *Cyclospora* frequently finds its way into water and then can be transferred to foods. It can also be transferred to foods during handling.

Prevention—Prevention against *Cyclospora* is similar to other parasites and viruses. Avoid contact with contaminated foods, water, or food employees. Good sanitation, good personal hygiene, and purchasing ready-to-eat foods from an inspected and approved supplier are also very important.

Cryptosporidium parvum; Giardia lamblia

CAUSATIVE AGENT	TYPE OF ILLNESS	SYMPTOMS ONSET	COMMON FOODS	PREVENTION
Cryptosporidium parvum	Parasitic infection	Severe watery diarrhea within 1 week of ingestion	Contaminated water, food contaminated by infected food employees	Use potable water supply; practice good personal hygiene and handwashing
Giardia lamblia	Parasitic infection	Diarrhea within 1 week of contact	Contaminated water	Potable water supply; good personal hygiene and handwashing

Description—*Cryptosporidium parvum* and *Giardia lamblia* are single-celled microorganisms called protozoa. *Cryptosporidium* is found in water that has been contaminated with cow feces. *Giardia* is found in the feces of wild animals, domestic pets, and infected persons. Both types of organisms can cause foodborne infection. These organisms are important sources of nonbacterial diarrhea in the United States.

Transmission in Foods—These parasites could occur, theoretically, on any food touched by a contaminated food employee. They are primarily transmitted by a water supply that is contaminated with feces and by fecal contamination of food and food-contact surfaces.

Prevention—Provide a potable water supply in the retail food establishment and handle foods carefully to prevent contamination and cross contamination. Food employees must practice good personal hygiene and wash hands thoroughly before working with food and after going to the toilet.

Toxoplasma gondii

CAUSATIVE AGENT	TYPE OF ILLNESS	SYMPTOMS ONSET	COMMON FOODS	PREVENTION
Toxoplasma gondii	Parasitic infection	Mild cases of the disease involve swollen lymph glands, fever, headache, and muscle aches. Severe cases may result in damage to the eye or brain (10 to 13 days)	Raw meats, raw vegetables, fruit	Good sanitation; reputable supplier; proper cooking

Description—Toxoplasmosis is a parasitic infection caused by *Toxoplasma gondii.* The parasite is common in warm-blooded animals including cats, rats, mice, pigs, cows, sheep, chickens, and birds.

Transmission in Foods—The domestic cat appears to be a major culprit in transmitting the parasite to humans and other animals. Food can be contaminated by contact with fecal material. Humans most often acquire infections by ingesting cysts in undercooked red meat. Unborn babies can catch this parasite from their mother if the mother is infected during pregnancy.

Prevention—Avoid eating raw and undercooked meat. Wash fruits and vegetables before eating and wash hands after handling raw meat and vegetables. Prevent cross contamination from raw foods to cooked or ready-to-eat foods by washing and sanitizing equipment and utensils.

Trichinella spiralis

CAUSATIVE AGENT	TYPE OF ILLNESS	SYMPTOMS ONSET	COMMON FOODS	PREVENTION
Trichinella spiralis	Parasitic infection from a nematode worm	Nausea, vomiting, diarrhea, sweating, muscle soreness (2 to 28 days)	Primarily undercooked pork products and wild game meats (bear, walrus)	Cook foods to the proper temperature throughout

Description—*Trichinella spiralis* is a foodborne roundworm that causes a parasitic infection. It must be eaten with the infected fleshy muscle of certain meat-eating animals to be transmitted to a new host.

Transmission in Foods—This parasite is frequently carried by meat-eating, scavenger animals. These animals are exposed to the parasite when they eat infected tissues from other animals and garbage that contains contaminated raw-meat scraps.

Prevention—Cook meat that could contain the parasite to the proper temperature. Make certain there are no signs of pink color in cooked pork products.

Problems Caused by Fungi

Yeast and molds are the types of fungi that have the greatest impact on food safety and quality. Both yeast and molds contribute to food spoilage. When yeast grows in a food, it leads to undesirable characteristics of a food including production of gases, acids, and alcohol. The food may taste or smell "sour" or the package may swell. Yeasts do not cause foodborne illness. In fact, yeasts are often used to produce fermented foods like beer, wine, and cheeses. Molds are also part of the fungi group. They usually grow best in foods that are low in moisture (bread or cheese), acidic (fruit juices), or high in sugar (jams and jellies). They often appear very colorful or "cotton-like" or "powdery" in appearance. Like yeasts, molds themselves do not cause foodborne illness. However, if they grow long enough on foods, they can produce a substance called a "mycotoxin." Mycotoxins can cause foodborne illness and some cancers. Mycotoxins are considered chemical hazards and are described in more detail later in this chapter.

Foodborne Illness Caused by Chemicals

Chemical hazards are usually classified as either naturally occurring or man-made chemicals. Naturally occurring chemicals include toxins that are produced by a biological organism. Man-made chemicals include substances that are added, intentionally or accidentally, to a food during processing. Figure 2.18 provides a summary of some of the more common naturally occurring and man-made chemicals.

Naturally Occurring:	Allergens in food
	Ciguatoxin
	Mycotoxin
	Scombrotoxin
	Shellfish toxins.
Man-made Chemicals:	Cleaning solutions
	Food additives
	Pesticides
	Heavy metals.

Figure 2.18 Types of Chemical Hazards in a Retail Food Establishment
(Source: *FDA Food Code*)

Naturally Occurring Chemicals

Allergens in Food

CAUSATIVE AGENT	TYPE OF ILLNESS	SYMPTOMS ONSET	COMMON FOODS	PREVENTION
Food Allergens	An allergic reaction usually involving the skin, mouth, digestive tract, or airways	Skin: hives, rashes, and itching Mouth: swelling and itching of the lips and tongue Digestive tract: vomiting and diarrhea Airways: difficulty breathing, wheezing	Foods that contain milk, egg, wheat, nuts and peanuts, fish, shellfish	Packaged and prepared foods must be properly labeled if they contain common food allergens so sensitive people can avoid them

Description—Between 5 and 8% of children and 1 to 2% of adults are allergic to certain chemicals in foods and food ingredients. These chemicals are commonly referred to as food allergens. A food allergen causes a person's immune system to "overreact." Some common symptoms of food allergies are hives; swelling of the lips, tongue, and mouth; difficulty breathing or wheezing; and vomiting, diarrhea, and cramps. These symptoms can occur in as little as five minutes after the food is eaten. In severe situations, a life-threatening allergic reaction called anaphylaxis can occur. Symptoms of anaphylaxis include itching and hives, swelling of the throat and difficulty breathing, lowered blood pressure and unconsciousness. Major food allergens are defined as the allergenic proteins from milk, egg, wheat proteins, peanuts, soy, tree nuts (including almonds, Brazil nuts, cashews, chestnuts, filberts/ hazelnuts, macadamia nuts, pecans, pine nuts, pistachios, and walnuts), fish, and shellfish (such as shrimp, lobster, crab, and crayfish). Major food allergens have been estimated to cause more than 90% of all food allergic reactions. Allergies have also been associated with sesame seeds, sunflower seeds, cottonseed (meal, not oil), poppy seed, molluscan shellfish, beans (other than green beans), peas, and lentils.

Transmission in Foods—Allergens can get into foods by way of cross contamination and accidental misformulation of products. Cross contamination of an ingredient or a product by a major food allergen from a different ingredient or a product may arise in storage and handling of raw materials, or during production due to residues in shared equipment, airborne dust, or the improper incorporation of re-work material without consideration of the allergen problem.

Accidental misformulation of products can result in the inclusion of a major food allergen (or any other ingredient) not prescribed in the product formulation.

Prevention—The only way for a person who is allergic to one of these foods to keep from having an allergic reaction is to avoid the food containing the allergen. In many cases, it doesn't take much of the food to produce a severe reaction. As little as half a peanut can cause a severe reaction in highly sensitive people. The responsibility of retailers in the case of prepackaged foods is to assure appropriate label warnings are given to potential purchasers (Figure 2.19). Retailers have special responsibilities in regard to foods that are sold from open containers (i.e., on delicatessen counters). When such foods contain a major food allergen, the warning should be clearly displayed by the foods in question. Moreover, staff must be trained to take great care to avoid cross contamination such as might occur when using the same ladle or other handling equipment for a food containing a major food allergen and one that does not contain it.

Figure 2.19 Label Warning of Food Allergen

Ciguatoxin

CAUSATIVE AGENT	TYPE OF ILLNESS	SYMPTOMS ONSET	COMMON FOODS	PREVENTION
Ciguatoxin	Fish toxin originating from toxic algae of tropical waters	Vertigo, hot/cold flashes, diarrhea, vomiting (15 min to 24 hrs)	Marine finfish including grouper, barracuda, snapper, jack, mackerel, triggerfish, reef fish	Purchase fish from a reputable supplier; cooking WILL NOT inactivate the toxin

Description—Ciguatoxin poisoning is an example of an intoxication caused by eating contaminated tropical reef fish.

Transmission in Foods—The toxin is found in tiny, free-swimming sea creatures called algae that live among certain coral reefs. When the toxic algae are eaten by small reef fish, it is stored in the flesh, skin, and organs. When the small reef fish are eaten by bigger fish such as barracuda, mackerel, mahi, bonito, jackfish, and snapper, the toxin accumulates in the flesh and skin of the larger fish. The contaminated fish are not affected by the toxin.

Prevention—The toxin is not destroyed by cooking; therefore, prevention can be very difficult. At the present time, there is no commercially known method to determine if ciguatoxin is present in a particular fish. Purchasing seafood from a reputable supplier is the best preventative measure.

Scombrotoxin

CAUSATIVE AGENT	TYPE OF ILLNESS	SYMPTOMS ONSET	COMMON FOODS	PREVENTION
Scombrotoxin	Seafood toxin originating from histamine-producing bacteria	Dizziness, burning feeling in the mouth, facial rash or hives, peppery taste in mouth, headache, itching, teary eyes, runny nose (1 to 30 min)	Tuna, mahi-mahi, bluefish, sardines, mackerel, anchovies, amberjack, abalone	Purchase fish from a reputable supplier; store fish at low temperatures to prevent growth of histamine-producing bacteria; toxin IS NOT inactivated by cooking

Description—Scombrotoxin, also called histamine poisoning, is caused by eating foods high in a chemical compound called histamine. Histamine is usually produced by certain bacteria when they decompose foods containing the protein histidine. Dark meat of fish has more histidine than other fish meat. Histamine is not inactivated by cooking.

Transmission in Foods—Over time, bacteria inherent to a particular food can break down histidine and cause the production of histamine. Leaving fish out at room temperature usually results in histamine production.

Prevention—Purchase seafood from a reputable supplier. Store "fresh" seafood at temperatures between 32°F (0°C) and 39°F (4°C), and do not accept seafood suspected of being thawed and refrozen or temperature abused in some way.

Shellfish Toxins—PSP, DSP, DAP, NSP

CAUSATIVE AGENT	TYPE OF ILLNESS	SYMPTOMS ONSET	COMMON FOODS	PREVENTION
Shellfish toxins: PSP, DSP, DAP, NSP	Intoxication	Numbness of lips, tongue, arms, legs, neck; lack of muscle coordination (10 to 60 min)	Contaminated mussels, clams, oysters, scallops	Purchase from a reputable supplier

Description—There are several shellfish toxins that have been associated with foodborne illness. The most common illnesses are Paralytic Shellfish Poisoning (PSP), Diarrhetic Shellfish Poisoning (DSP), Domoic Acid Poisoning (DAP), and Neurotoxic Shellfish Poisoning (NSP). The toxins are produced by certain algae called dinoflagellates. When filter-feeding shellfish, such as mussels, clams, oysters, and scallops, feed on the toxic algae, they accumulate the toxins in their internal organs and become toxic to humans. The amount of toxin in the shellfish depends on the amount of toxic algae (dinoflagellates) in the water and the amount of water filtered by the shellfish.

Transmission in Foods—Most cases of seafood toxins are caused by contaminated shellfish that have been harvested by sport fishermen or poached from polluted waters. Commercially harvested shellfish are rarely involved in foodborne disease outbreaks because health agencies monitor the level of toxin in shellfish during high-risk periods (May to

October). Toxin levels in the shellfish are determined in public health laboratories. When the amount of toxin in the shellfish exceeds safe limits, sale of the shellfish from that harvesting area is prohibited.

Prevention—Purchase shellfish only from a reputable supplier. Avoid buying shellfish that have been harvested by sport fishermen or poached from polluted waters.

Mycotoxins

CAUSATIVE AGENT	TYPE OF ILLNESS	SYMPTOMS ONSET	COMMON FOODS	PREVENTION
Mycotoxins	Intoxication	1)Acute onset-hemorrhage, fluid buildup 2) Chronic onset-cancer from small doses over time	Moldy grains, corn, corn products, peanuts, pecans, walnuts, milk	Purchase food from a reputable supplier; keep grains and nuts dry; protect products from humidity; visually inspect foods for signs of mold growth

Description—Mycotic organisms or fungi are molds, yeasts, and mushrooms, some of which are capable of causing foodborne illness. Fungi are larger than bacteria, and they usually prefer foods high in sugar or starches. Molds and yeasts withstand more extreme conditions (more acidic foods, lower A_w foods) than bacteria can. Most molds and yeasts are spoilage organisms, which cause foods to deteriorate. However, some types of fungi produce toxic chemicals called mycotoxins. Many mycotoxins have been shown to cause cancer. An important foodborne mycotoxin, called aflatoxin, is produced by *Aspergillus spp.* molds. Many mycotoxins are not destroyed by cooking.

Transmission in Foods—Foodborne fungi are important because they can produce chemical compounds called mycotoxins that have been linked to cancer.

Prevention—Purchase food from a reputable supplier. Keep grains and nuts dry and protected from humidity. Visually inspect foods for any signs of mold growth.

Added Man-Made Chemicals

There is an extensive list of chemicals added to foods that may pose a potential health risk. Intentionally added chemicals may include food additives, food preservatives, and pesticides. Pesticides leave residues on fruits and vegetables and can usually be removed by a vigorous washing procedure. Non-intentionally added chemicals may include contamination by chemicals such as cleaning and sanitary supplies. Also, chemicals from containers or food-contact surfaces of inferior metal that are misused may lead to heavy-metal or inferior-metal poisoning (cadmium, copper, lead, galvanized metals, etc.).

Foodborne Illness Caused by Physical Hazards

Physical hazards are foreign objects in food that can cause illness and injury (Figure 2.20). They include items such as fragments of glass from broken glasses, metal shavings from dull can openers, unfrilled toothpicks that may contaminate sandwiches, human hair and jewelry, or bandages that may accidentally be lost by a food handler and enter food. Stones, rocks, or wood particles may contaminate raw fruits and vegetables, rice, beans, and other grain products.

Figure 2.20 Common Physical Hazards in a Food Establishment
(Source: *FDA Food Code*)

Physical hazards commonly result from accidental contamination and poor food-handling practices that can occur at various points in the food chain from harvest to consumer. To prevent physical hazards, wash raw fruits and vegetables thoroughly and visually inspect foods that cannot be washed (such as ground beef). Food employees must be taught to handle food safely to prevent contamination by unwanted foreign objects such as glass fragments and metal shavings. Finally, food employees should not wear jewelry when involved in the production of food, except for a plain wedding band.

In this chapter you learned about the many agents that can cause foodborne illness. The summary of disease-causing agents presented in Appendix B provides a quick reference to foodborne illnesses. In the busy work world of a retail food establishment manager, this reference can save you time when searching for such information.

Problems in Other Countries Related to Food

Mad Cow Disease or Bovine Spongiform Encephalopathy (BSE) is a fatal brain disorder that occurs in cattle and is caused by some unknown agent. There seems to be a connection between animal feed made from the parts of sheep that carry the organisms called scrapie. Use of animal feed made from such ingredients has been banned in the United States.

The connection between BSE and humans was uncovered in Great Britain in the 1990s when several young people died of a brain disorder, a new variation of a rare problem, Creutzfeldt-Jakob disease (CJD) that typically strikes the elderly.

Efforts of the Food and Drug Administration (FDA), U.S. Department of Agriculture (USDA), the Centers for Disease Control and Prevention (CDC), and other federal organizations, including state regulatory and health agencies, have kept the disease from occurring in this country. Advisories are regularly issued to those traveling to Great Britain and Europe about the safety of eating beef and beef products (FDA Backgrounder, 2001).

Foot- (hoof) and-mouth disease is a highly contagious viral infection of cattle, sheep, goats, deer, and other cloven-hoofed animals. It causes blisters on the mouth, teats, and soft tissues of the animal's feet and mouth. The animals rarely recover. The danger from this disease is to animals, not humans. Therefore, foot-and-mouth disease is not classified as a foodborne illness.

An outbreak of foot-and-mouth disease occurred in the United Kingdom (England, Scotland, Wales, and Northern Ireland) and Europe in 2001. Affected areas were placed under quarantine, and a large number of animals were destroyed to contain the infection. This outbreak had a negative economic impact on the farmers whose herds were infected.

Summary

Biological hazards such as bacteria, viruses, and parasites continue to cause problems. An increase in foodborne illness associated with fresh produce was described at the beginning of this chapter. This change is due to several factors. People eat more fruits and vegetables because of the known health benefits. An increase in the number of people in the at-risk population exposes more individuals to foodborne illness. The number and types of hazards also seem to be on the increase. Shiga toxin-producing *Escherichia coli* was not a significant problem until the last decade. Now, this organism seems to be in the headlines nearly every day. Retail food establishment managers need to be aware of the change in eating habits, marketing patterns, and emerging pathogens in our food supply.

With fruit and vegetable products, we learned potentially hazardous foods are not the only problems. Apple cider, for example, has a pH well below 4.6, not a potentially hazardous food. However, certain microbes like Shiga toxin-producing *Escherichia coli* can survive and still cause illness. There are some preventative measures described in later chapters of this book that can be taken to reduce the risk of fresh produce. Wash fresh produce thoroughly to lessen contamination. Use self-inspections and audits for all products from the supplier of the fruits and vegetables. The procedures used by the supplier become very important. With other products, some retail food operations choose not to take the risk; they simply do not offer them for sale. Unpasteurized apple cider is a good example. Products can also be labeled to inform the customer of the potential risk. The FDA now requires a label on unpasteurized apple cider stating: "WARNING: This product has not been pasteurized and, therefore, may contain harmful bacteria that can cause serious illness in children, the elderly, and persons with weakened immune systems."

There are many foodborne hazards a retail food establishment may encounter. They are classified as biological, chemical, or physical hazards. These hazards differ depending on the type of food and method of preparation involved. Retail food establishments are typically toward the end of the food production chain. This is where foods are prepared or sold for consumer preparation. Therefore, it is very important to control and prevent foodborne hazards as much as possible to reduce the risk of foodborne illness associated with your establishment. Control and prevention of foodborne hazards in a retail food establishment start with understanding the different types of foodborne hazards. The next step is to understand how to control foodborne hazards with time/temperature control, good personal hygiene, cleaning and sanitation, and prevention of cross contamination. Prevention, using these four approaches, will be the focus of Chapter 3.

Discussion Questions (Short Answer)

1. Briefly explain the difference between an infection, an intoxication, and a toxin-mediated infection. Give an example of each.
2. What groups of people have a greater risk of acquiring foodborne illness?
3. Identify the three categories of hazards that can cause foodborne illness and give examples of each.
4. What types of biological hazards should be of greatest concern to retail food establishment managers? Why?
5. What does FATTOM represent?
6. At what pH and water activity (A_w) levels do disease-causing bacteria grow best?
7. What is the food temperature danger zone?
8. What is a bacterial spore? Why is it important for the retail food establishment manager to understand?
9. What are potentially hazardous foods? What three characteristics do they share?
10. What are viruses? What are parasites?

Quiz 2 (Multiple Choice)

1. Bacteria are a common cause of foodborne disease in a retail food establishment because:
   a. Under ideal conditions, they can grow very rapidly.
   b. Bacteria are found naturally in many foods.
   c. Bacteria can be easily transferred from one source to another.
   d. All of the above.
2. Most of the bacteria that cause foodborne illness grow:
   a. With or without oxygen at an ideal temperature of 98.6° F (39°C).
   b. Only without oxygen at an ideal temperature of 110°F (43°C).
   c. Only with oxygen at an ideal temperature of 110°F (43°C).
   d. Only without oxygen at an ideal temperature of 98.6°F (39°C).
3. Which of the following organisms are most likely to cause a foodborne disease outbreak?
   a. *Salmonella spp.* and *Campylobacter jejuni.*
   b. *Cryptosporidium parvum* and *Giardia lamblia.*
   c. Hepatitis A and Norwalk virus group.
   d. *Anisakis spp.* and *Trichinella spiralis.*
4. Bacteria grow best within a narrow temperature range called the temperature danger zone. The temperature danger zone is between:
   a. 0°F (-18°C) and 220°F (104°C).
   b. 0°F (-18°C) and 140°F (60°C).
   c. 41°F (5°C) and 140°F (60°C).
   d. 41°F (5°C) and 220°F (104°C).
5. Bacteria that cause foodborne illness will only grow on foods that have a pH at ____ or above and a water activity (A_w) above ______.
   a. 3.2; .85.
   b. 4.6; .85.
   c. 6.5; .80.
   d. 8.0; .70.

6. Which of the following bacteria produce a toxin that is most likely to cause death if consumed?
   a. *Campylobacter jejuni.*
   b. *Clostridium botulinum.*
   c. Shiga toxin-producing *Escherichia coli.*
   d. *Listeria monocytogenes.*
7. Some bacteria form spores to help them:
   a. Reproduce and grow in numbers.
   b. Move more easily from one location to another.
   c. Survive adverse environmental conditions.
   d. Grow in high acid foods.
8. Which of the following is a histamine poisoning?
   a. Ciguatoxin.
   b. Scombrotoxin.
   c. Mycotoxin.
   d. Paralytic Shellfish Poisoning (PSP).
9. Which of the following is not considered a potentially hazardous food group?
   a. Red meats.
   b. Fish and shellfish.
   c. Poultry and eggs.
   d. Dried grains and spices.
10. The most effective way to control the growth of bacteria in a retail food establishment is by controlling:
    a. Time and temperature.
    b. pH and oxygen conditions.
    c. Temperature and water activity.
    d. Time and food availability.

Answers to the multiple choice questions are provided in Appendix A.

References/Suggested Readings

Adams, M. R. and M. O. Moss (2000). *Food Microbiology*. Royal Society of Chemistry. London, England.

Banwart, G. J. (1989). *Basic Food Microbiology* (2nd Edition). Van Nostrand Reinhold: New York.

Chin, James (Ed.) (2000). *Control Of Communicable Diseases in Man* (17th Edition). American Public Health Association. Washington, DC.

Doyle, Michael P., Larry R. Beuchat, and Thomas J. Montville (1997). *Food Microbiology: Fundamentals and Frontiers.* American Society for Microbiology. Washington, D.C.

Food and Drug Administration (2001). *2001 Food Code.* U.S. Public Health Service. Washington, DC.

Food Marketing Institute (2002). *Hepatitis A Information Guide.* Food Marketing Institute, 655 15th. Street, N.W. Washington, DC.

Jay, James J. (2000). *Modern Food Microbiology* (6th Edition). Aspen Publishers. Gaithersburg, MD.

Longreè, K., and G. Armbruster (1996). *Quantity Food Sanitation.* Macmillan. New York, NY.

McSwane, D., Rue, N., Linton, R. (2002). *Essentials of Food Safety and Sanitation* (3rd Edition). Prentice Hall, Upper Saddle River, NJ.

Suggested Web Sites

Gateway to Government Food Safety Information
www.foodsafety.gov

Centers for Disease Control and Prevention (CDC)
www.cdc.gov

Bad Bug Book
vm.cfsan.fda.gov/~mow/intro.html

USDA/FDA Food and Nutrition Information Center
www.nal.usda.gov/fnic/

The Food Allergy Network
www.foodallergy.org

The Food Marketing Institute
www.fmi.org

United States Department of Agriculture (USDA)
www.usda.gov

United States Food and Drug Administration (FDA)
www.fda.gov

United States Environmental Protection Agency (EPA)
www.epa.gov

Partnership for Food Safety Education
www.fightbac.org

3 Preventing Foodborne Illness

Hepatitis A Outbreak Traced to Seafood Department of Local Retail Food Establishment

A local health department received a telephone call from a man who wanted an immune globulin vaccine after being exposed to Hepatitis A by his roommate. A lab test showed the man was immune globulin M positive for Hepatitis A, which demonstrated he had been exposed during the past three months. An interview with the victim revealed he worked in the seafood department of a local retail food establishment.

An investigation of the retail food establishment revealed two other food handlers who had been experiencing symptoms of Hepatitis A. Both employees tested positive for the disease. All the employees were responsible for preparing and handling raw and ready-to-eat foods in the department. The employees admitted they sometimes did not wear gloves or use utensils when they handled ready-to-eat foods. They also admitted they did not always wash their hands after going to the toilet.

Hepatitis A virus is shed in the feces of infected people. An employee's hands can become contaminated when he or she goes to the toilet. Food handlers can spread the virus even when they don't have symptoms of the disease.

What must be done to protect customers of the seafood department from the Hepatitis A virus?

Learning Objectives

After reading this chapter, you should be able to:

- ▲ Identify potential problems related to temperature and time abuse of foods.
- ▲ Describe how to properly measure and maintain safe food temperatures.
- ▲ Identify potential problems related to a food employee's personal hygiene practices.
- ▲ Outline a plan to teach food employees necessary personal hygiene habits.
- ▲ Identify potential problems related to cross contamination of food.
- ▲ Discuss procedures and methods to prevent cross contamination.

Essential Terms

Calibrate	**Food temperature-measuring device**
Centers for Disease Control and Prevention (CDC)	**Personal hygiene**
Cross contamination	**Temperature abuse**
Dial face bi-metal thermometer	**Thermocouple**
Digital thermometer	**T-Stick type melt devices**

Factors That Contribute to Foodborne Illness

The Centers for Disease Control and Prevention (CDC) is an agency of the U.S. Government responsible for collecting statistics about diseases including foodborne illness. CDC statistics show most outbreaks of foodborne disease occur because food is mishandled. The leading factors that cause foodborne illness, as identified by the CDC, are presented in Figure 3.1.

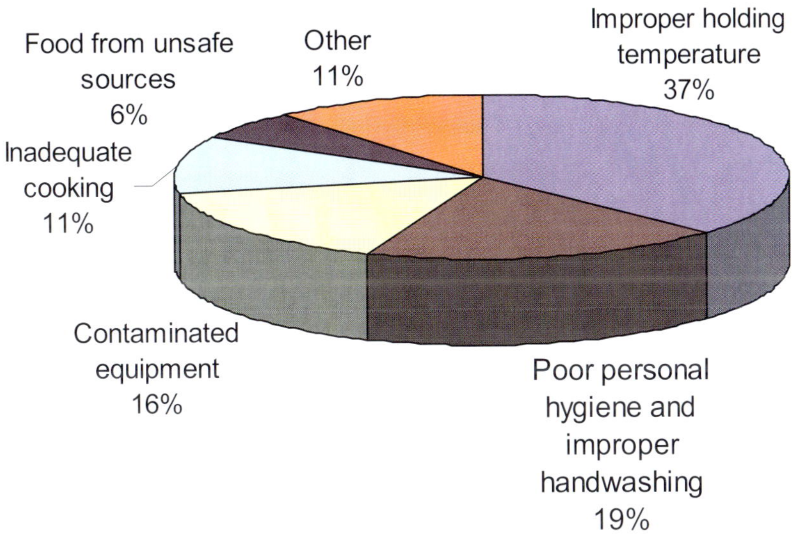

Figure 3.1 Factors Leading to Foodborne Illness (ranked by percentage of outbreaks)

Almost all foodborne illnesses are linked to:

- 37% - Improper holding temperature
- 19% - Poor personal hygiene and improper handwashing
- 16% - Contaminated equipment
- 11% - Inadequate cooking
- 6% - Food from unsafe sources
- 11% - Other.

It is extremely important you recognize these as major contributors to food contamination. The four factors shown in Figure 3.2 are the most important elements of food safety control that will be highlighted throughout this book. This chapter introduces the prevention strategies necessary for safe food management.

Figure 3.2 The Most Important Elements of Food Safety Control

What is Time and Temperature Abuse?

Controlling time and temperature are perhaps the most critical ways to ensure food safety. Most cases of foodborne illness can in some way be linked to temperature abuse. The term temperature abuse is used to describe situations where foods are:

- ▲ Exposed to temperatures in the temperature danger zone long enough to allow growth of harmful microorganisms
- ▲ Not cooked or reheated sufficiently to destroy harmful microorganisms that may be present in food.

In Chapter 2, *Hazards to Food Safety,* you learned harmful microbes could grow in potentially hazardous foods when temperatures are between 41°F (5°C) and 140° F (60°C).

An important rule to remember for avoiding temperature abuse is **Keep Hot Foods Hot**, **Keep Cold Foods Cold**, or **Don't Keep the Food at All**.

To prevent harmful microbes from growing, the temperature inside the core of a hot food item must be maintained above 140°F (60°C) as illustrated in Figure 3.3. The high temperatures reached during cooking destroy microbes. However, toxins produced by microbes may not be destroyed by normal cooking processes and could cause foodborne illness.

Figure 3.3 Temperature Danger Zone

Keep cold food temperatures below the temperature danger zone (41°F, 5°C) to prevent most harmful microbes from growing. Bacteria, such as *Listeria monocytogenes*, that can grow at these lower temperatures do so very slowly.

There are unavoidable situations during food production when foods must pass through the temperature danger zone such as:

- ▲ Cooking
- ▲ Cooling
- ▲ Reheating
- ▲ Food handling (slicing, mixing, sandwich assembly, etc.).

During these activities, you must minimize the amount of time foods are in the temperature danger zone to control microbial growth. ***When it is necessary for a food to pass through the temperature danger zone, do it as quickly as possible. In addition, foods should pass through the danger zone as***

few times as possible. As a rule, hot foods should be cooled and reheated as few times as possible.

Cooking foods to make them more enjoyable to eat began with the discovery of fire. Heating improves texture and flavor and also destroys harmful microorganisms. As you learned in Chapter 2, many raw foods naturally contain harmful microbes or can become contaminated during handling. When you cook and reheat foods properly, microbes are reduced to safe levels or are destroyed. *Cooking and reheating are two very important processes for safe food management.*

How to Measure Food Temperatures

Maintaining safe food temperatures is an essential and effective part of preventing foodborne illness. You must know how and when to measure food temperatures correctly. Thermometers, thermocouples, or other devices are used to measure the temperature of stored, cooked, hot-held, cold-held, and reheated foods. A temperature-measuring device with a small diameter probe must be available to measure the temperature of thin foods such as meat patties and fish filets.

Temperature-measuring devices typically measure food temperatures in degrees Fahrenheit (denoted as °F), degrees Celsius (denoted as °C), or both. It is important you know which temperature scale(s) your temperature-measuring device is measuring. Throughout this book, temperatures will be provided in both Fahrenheit and Celsius to acquaint you with both temperature scales. For example, the temperature requirement for cold-holding potentially hazardous foods will be noted as "41°F (5°C) or below." This simply means the food should be kept at or below 41 degrees Fahrenheit or 5 degrees Celsius.

Thermometer	Features/Uses:
Dial face, metal stem type (bi-metallic) 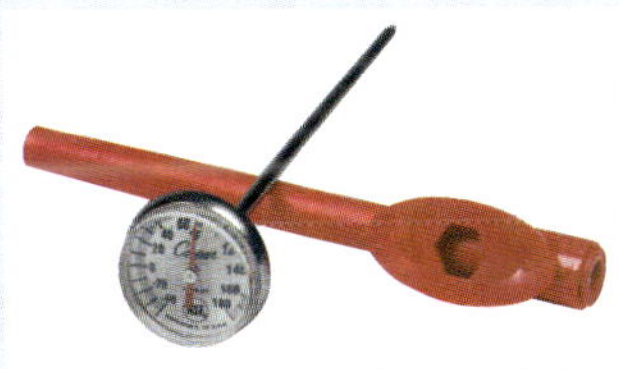 Courtesy of Cooper Instrument Corp.	Used to measure internal food temperatures at every stage of food preparation. Dial face – most common type of thermometer used. • Measures temperatures ranging from 0°F (-18°C) to 220°F (104°C) with 2°F increments • To ensure accurate measurements, the stem of a bi-metallic thermometer must be inserted at least 2 inches into the food item being measured
Digital 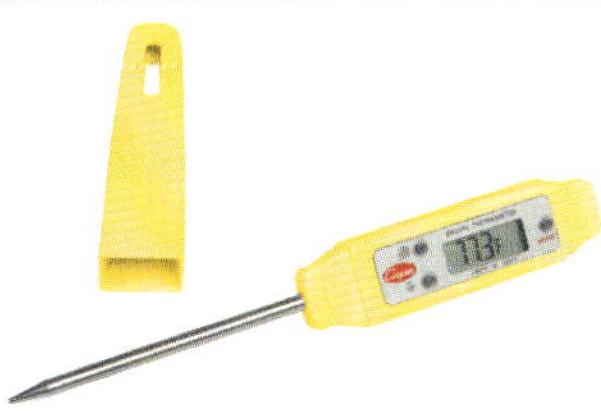 Courtesy of Cooper Instrument Corp.	Displays the temperature numerically. • Measures a wider range of temperatures than a dial face thermometer
Thermocouple  Courtesy of Cooper Instrument Corp.	Provides a digital readout of the temperature and has a variety of interchangeable probes for different applications.

Figure 3.4 Temperature-Measuring Devices

Thermometer	Features/Uses:
Infrared 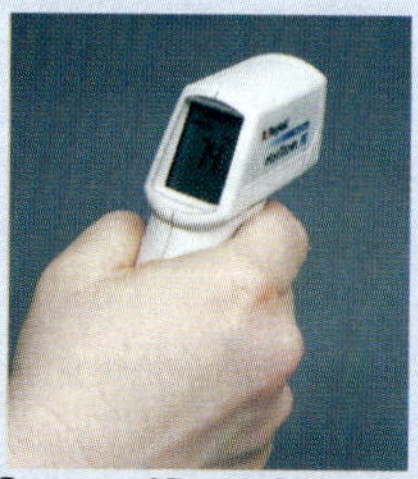 Courtesy of Raytek Corporation	Measures the surface temperature of food without actually touching the food. • Can measure many different products without cross contamination • Check the accuracy frequently • Potential for thermal shock — (i.e. needs about 20 minutes to adjust between hot and cold temperatures)
T-Sticks (melt device) 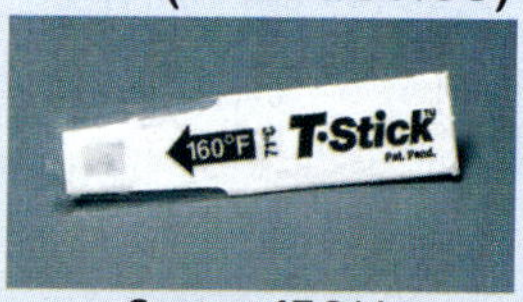 Courtesy of T-Stick	Single–use disposable thermometer measures only one temperature. • Change color when they reach the indicated temperature • Used to monitor product temperatures and sanitizing temperature in dishwashing machines
Built-in 	Refrigerated and frozen food cases contain built-in thermometers to check temperatures for food storage.
Maximum Registering (holding) 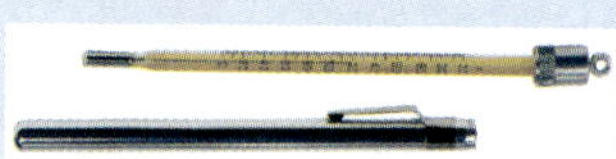 Courtesy of Cooper Instrument Corp.	Measures the temperature of hot water used to sanitize dishware and utensils in mechanical warewashing machines. This device is becoming less popular in retail food establishments because it contains mercury which can be a contaminant of food and the general environment.

Figure 3.4 Temperature-Measuring Devices (Cont.)

Thermometer Guidelines	• Food temperature-measuring devices scaled only in Celsius or dually scaled in Celsius and Fahrenheit must be accurate to ±1.8°F (±1°C). Food temperature-measuring devices scaled in Fahrenheit only must be accurate to ±2°F. • Mercury filled and glass thermometers should not be used in retail food establishments. • Clean and sanitize thermometers properly to avoid contaminating food that is being tested. This is very important when testing raw and then ready-to-eat food items. To clean and sanitize a food thermometer, wipe off any food particles, place the stem or probe in sanitizing solution for at least 5 seconds, then air dry. • When monitoring only raw foods, or only cooked foods being held at 140°F (60°C), wipe the stem of the thermometer with an alcohol swab between measurements.

Figure 3.4 Temperature-Measuring Devices (Cont.)

When and How to Calibrate Thermometers

It is very important thermometers are accurate in measuring the temperature of foods. Thermometers must be calibrated on a regular basis. **Calibrate** means to compare and adjust with a known standard.

The two temperature-measuring devices most commonly used in retail food establishments are the dial face thermometer and the digital thermometer. It is important to calibrate dial face metal stem-type (bi-metal) thermometers:

- ▲ Before their first use
- ▲ At regular intervals
- ▲ If dropped or otherwise damaged
- ▲ If used to measure extreme temperatures

▲ Whenever accuracy is in question.

Most dial face thermometers have a calibration nut just behind the dial (Figure 3.5). These devices are calibrated using a boiling point or ice point method. Specific directions for calibrating this type of thermometer are described in the following section. Check manufacturer's directions for any questions you may have.

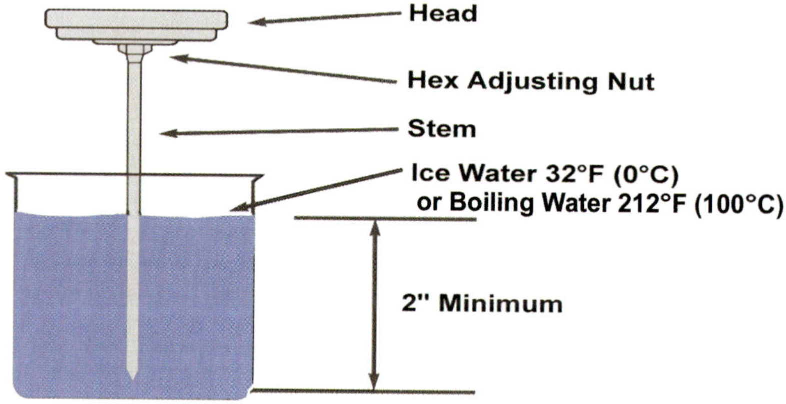

Figure 3.5 Calibrating a Dial Face Thermometer

(Source: Cooper Instrument Corporation)

Boiling Point Method

Immerse at least the first 2 inches of the stem from the tip (the sensing part of the probe) into boiling water and adjust the needle to 212°F (100°C). At higher altitudes, the temperature of the boiling point will vary. Consult your local health department if you have any questions about the boiling point of water in your area.

Ice Point Method

Insert the probe into a cup of crushed ice. Add enough cold water to the cup to remove any air pockets that might remain. Wait until the temperature stabilizes in the ice mixture and adjust the needle to 32°F (0°C).

Measuring Food Temperature

The sensing portion of a food thermometer is near the end of the stem or probe. On the bi-metal thermometer, the sensing portion extends from the tip up to the "dimple" mark on the stem. The dimple is typically about 1 inch from the tip of the probe. An average of the temperature is measured over this distance. The sensing portion for digital and thermocouple thermometers is closer to the tip of the probe.

Accurate readings are only possible when the sensing portion of the temperature-measuring device is inserted deeply into the food. For bi-metal thermometers, immerse the needle tip at least 2 inches into the food to be measured. For digital and thermocouple thermometers, the tip must be inserted 1 inch or more. Always insert the sensing element of the thermometer into the center or thickest part of the food. When possible, stir the food before measuring the temperature. Always wait for the temperature reading to stabilize. [See Figure 3.6(a) and (b).]

The approximate temperature of packaged foods can be measured accurately without opening the package. Place the stem or probe of the thermometer between two packages of food or fold the package around the stem or probe to make good contact with the packaging [Figure 3.6(c) and (d)]. For some prepackaged foods, it may be best to open the package or container and measure the temperature of the food [Figure 3.6(e)]. To accurately and safely measure food temperatures, be sure to:

- ▲ Use an approved temperature-measuring device that measures temperature from 0°F (-18°C) to 220°F (104°C)
- ▲ Locate the sensing portion of the measuring device
- ▲ Calibrate the measuring device using the ice or boiling point method
- ▲ Clean and sanitize the probe of the temperature-measuring device according to procedure
- ▲ Measure the internal temperature of the food by inserting the probe into the center or thickest part of the item.

a. In a salad bar

b. For rotisserie chicken

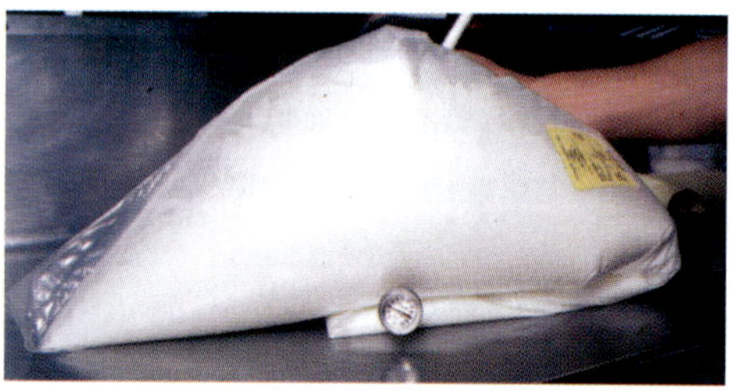

c. For bulk storage of milk

d. For prepackaged meats

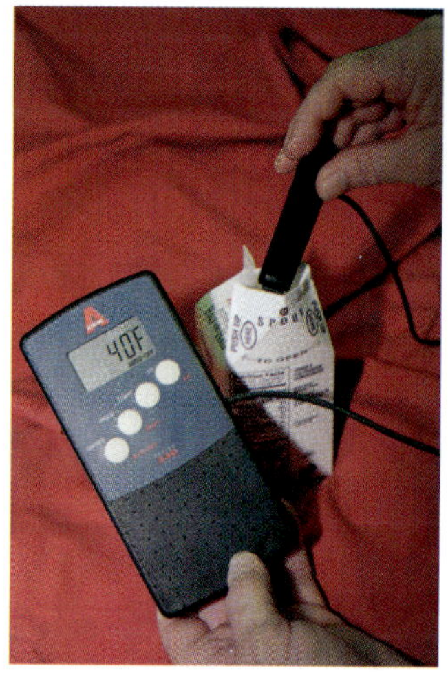

e. For single-use containers

Figure 3.6 Measuring Food Temperature

Preventing Temperature Abuse

Controlling the temperature of potentially hazardous food is important in almost all stages of food handling. Thus, measuring temperatures of potentially hazardous food is a very important responsibility for food handlers. The following table (Figure 3.7) lists safe temperature guidelines with the rationale for each step. Each of these guidelines will be discussed throughout the remainder of this book.

Steps in the Flow of Food	Safe Temperature Guidelines	Rationale for Temperature Guidelines
Receiving and Storing Frozen Foods	Foods should be frozen solidly and maintained frozen at all times.	Proper freezing of foods helps to maintain food quality and prevents the growth of spoilage and harmful microorganisms.
Receiving and Storing Refrigerated Foods	Foods should be received and stored so food is always at or below 41°F (5°C). Raw shell eggs may be received at 45°F (7°C) ambient air temperature or below.	Receiving and storing foods below 41°F (5°C) prevent or slow the growth of harmful microorganisms.
Cooking Foods	Different foods, and the methods by which they are cooked, require different end point temperatures to be safe. The range of safe cooking temperatures can vary from 145°F (63°C) to 165°F (74°C). Beef roasts maybe cooked at 130°F (54°C) for rare doneness.	Proper cooking destroys harmful microorganisms that may be present in the food.
Cooling Foods	During cooling, food must be cooled from 140°F (60°C) to 70°F (21°C) within 2 hours and from 140°F (60°C) to 41°F (5°C) within 6 hours.	Proper cooling prevents the conversion of spores to vegetative bacterial cells and the subsequent growth of vegetative cells.

Figure 3.7 Time and Temperature Controls for Potentially Hazardous Food

Steps in the Flow of Food	Safe Temperature Guidelines	Rationale for Temperature Guidelines
Reheating Foods	All reheated foods must be reheated to at least 165°F (74°C) within 2 hours. Ready-to-eat food that has been commercially processed and is in an intact package can be reheated to 140°F (60°) within 2 hours.	Proper reheating destroys harmful bacteria that may be present in foods.
Hot-holding Foods	All foods must be cooked to a safe temperature and then held at greater than 140°F (60°C).	Proper holding of food prevents the growth of harmful bacteria.
Cold-Holding Foods	All foods that are held and served cold must be held at 41°F (5°C) or below.	Holding cold foods below 41°F (5°C) prevents or slows the growth of harmful microorganisms.
Thawing Foods	Thawing may be done in a refrigerator at 41°F (5°C) or less, in a microwave oven and then immediately cooked, or under cool running water at 70°F (21°C).	Proper thawing prevents or reduces the growth of harmful bacteria.
Food Preparation	During food preparation, food should only be in the temperature danger zone [between 41°F (5°C) and 140° F (60°C)] for a maximum total time of 4 hours.	Maintaining foods between 41° (5°C) and 140°F (60°C) for no more than 4 hours limits the number of microorganisms that can grow.

Figure 3.7 Time and Temperature Controls for Potentially Hazardous Food (Cont.)

Keep Cold Foods Cold and Hot Foods Hot!

Frozen foods should be kept solidly frozen until they are ready to be used. Freezing helps to retain product quality. Proper frozen food temperatures do not permit disease-causing and spoilage microorganisms to grow. Cold temperatures also help preserve the color and flavor characteristics of the food. Frozen foods can be stored for long periods of time without losing their wholesomeness and quality.

Refrigerated foods are held cold, not frozen. Cold foods should be maintained at 41°F (5°C) or below. Do not forget that some harmful bacteria and many spoilage bacteria can grow at temperatures below 41°F (5°C), although their growth is very slow. By keeping cold foods at 41°F (5°C) or below, you can reduce the growth of most harmful microorganisms and extend the shelf life of the product. For maximum quality and freshness, hold cold foods for the shortest amount of time possible.

The application of heat, as in cooking, reheating, and hot-holding of foods, is another method used to preserve food. To destroy harmful bacteria, heat food to proper temperatures. Established safe cooking temperatures are based on the type of food and the method used to heat the product. Cooked foods that have been cooled and then reheated must be maintained at 140°F (60°C) or above until used. You must keep foods hot to stop growth of harmful bacteria.

There are times during food production when foods must be in the temperature danger zone. Recognize the time spent in the temperature danger zone should be minimal for potentially hazardous items (Figure 3.8).

KEY CONCEPT

Improper holding temperature of foods is the No. 1 contributing factor that leads to foodborne illness. Spores of certain bacteria like *Clostridium botulinum*, *Clostridium perfringens*, and *Bacillus cereus* can survive cooking temperatures. Remember, if spores survive and are exposed to ideal conditions, they can become vegetative cells and begin to grow in foods (Figure 2.2).

KEY CONCEPT

The *FDA Food Code* contains cooling guidelines that permit foods to be in the temperature danger zone for a total of six hours. The *FDA Food Code* specifically states foods must be cooled from 140°F (60°C) to 70°F (21°C) in two hours, and from 140°F (60°C) to 41°F (5°C) or less within six hours.

KEY CONCEPT

To destroy many of the bacteria that may have grown during the cooling process, reheat foods to 165°F (74°C) within two hours to prevent the number of organisms from reaching levels that can cause foodborne illness.

The preferred method for thawing foods is in a refrigerator at 41°F (5°C) or below. This prevents the food from entering the food temperature danger zone. Other acceptable methods for thawing include placing the food under cool running water at 70° (21°C) or using a microwave oven when the food is going directly to the cooking process. Proper thawing reduces the chances for bacterial growth, especially on the outer surfaces of food.

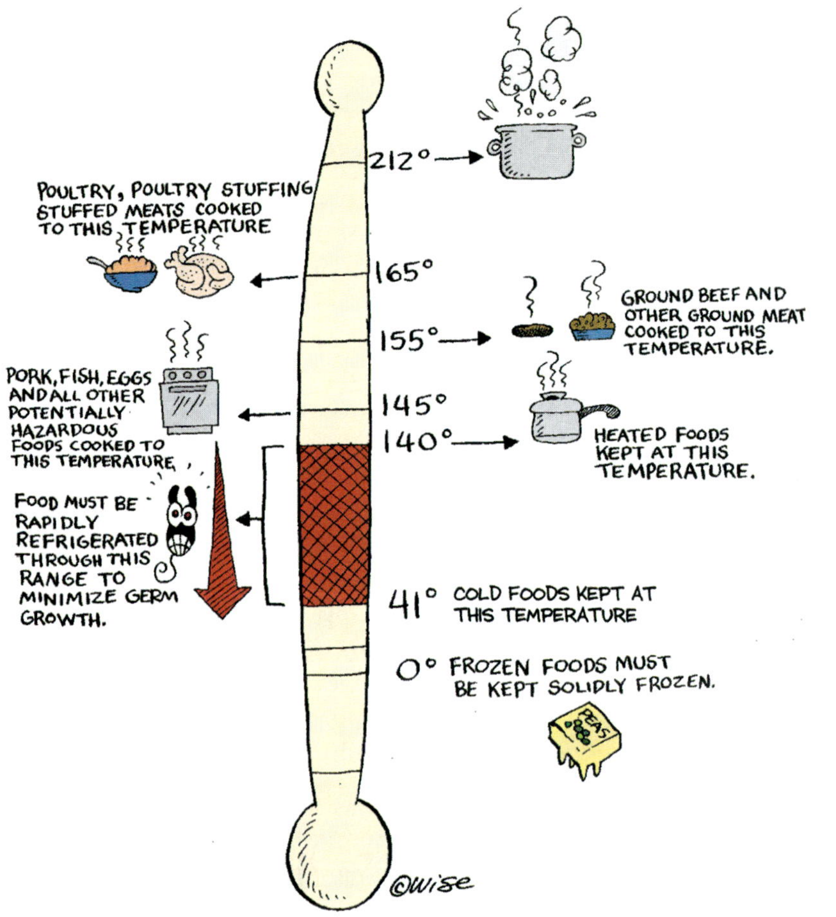

Figure 3.8 Keep it Hot, Keep it Cold, or Don't Keep It!

More detailed strategies for minimizing the amount of time a food is in the temperature danger zone during cooling, thawing, and food preparation will be presented in the next chapter.

The Importance of Handwashing and Good Personal Hygiene

The cleanliness and personal hygiene of food employees are extremely important. If a food employee is not clean, the food can become **contaminated**.

As you know, even healthy humans can be a source of harmful microorganisms. All of us carry bacteria on our hands. Therefore, **good personal hygiene is essential for those who handle foods**. Desirable behaviors include:

- ▲ Knowing when and how to properly wash hands
- ▲ Wearing clean clothing
- ▲ Maintaining good personal habits (bathing, washing and restraining hair, keeping fingernails short and clean; washing hands after using toilet; etc.)
- ▲ Maintaining good health and reporting when sick to avoid spreading possible infections.

Just think of all the things a food handler's hands touch during a typical workday. They may take out the trash, cover a sneeze, scratch an itch, or mop up a spill. Potentially harmful germs can contaminate their hands when they touch their face or skin, run their fingers through their hair or beard, use the toilet, or blow their nose. *Staphylococcus aureus,* Hepatitis A, and *Shigella spp.* are examples of pathogens that are commonly transferred to foods after contact with human hands.

Personnel involved in food handling must know how and when to wash their hands. Figure 3.9 provides a list of situations when handwashing is required.

Always wash hands and exposed portions of arms:

- Before food preparation
- After touching bare human body parts, except clean hands and clean exposed arms
- After using the toilet
- When returning to a work station
- After coughing, sneezing, using a handkerchief or tissue, using tobacco, eating, or drinking
- During food preparation as often as necessary to remove soil and contamination and to prevent cross contamination.
- Before putting on gloves for working with food
- After engaging in any activities that may contaminate hands (taking out the garbage, wiping counters or tables, handling cleaning chemicals, picking up dropped items, etc.)
- After caring for or touching service animals or aquatic animals.

Figure 3.9 Handwashing Guidelines

(Source: *FDA Food Code*)

Using approved cleaning compounds (soap or detergent), vigorously rub surfaces of fingers and fingertips, front and back of hands, wrists, and forearms (or surrogate prosthetic devices for hands or arms) for at least 20 seconds. Remember, soap, warm water, and friction are needed to adequately clean skin. A significant number of germs are removed by friction alone. A brush can be helpful when cleaning hands. However, the brush must be kept clean and sanitary.

Thoroughly rinse under clean warm running water and clean under and around fingernails and between fingers.

Dry hands using a single-service paper towel, an electric hand dryer, or clean section of continuous rolled cloth towel (if allowed in your area). Do not dry hands on your apron or a dishtowel. (See the handwashing procedure in Figure 3.10.)

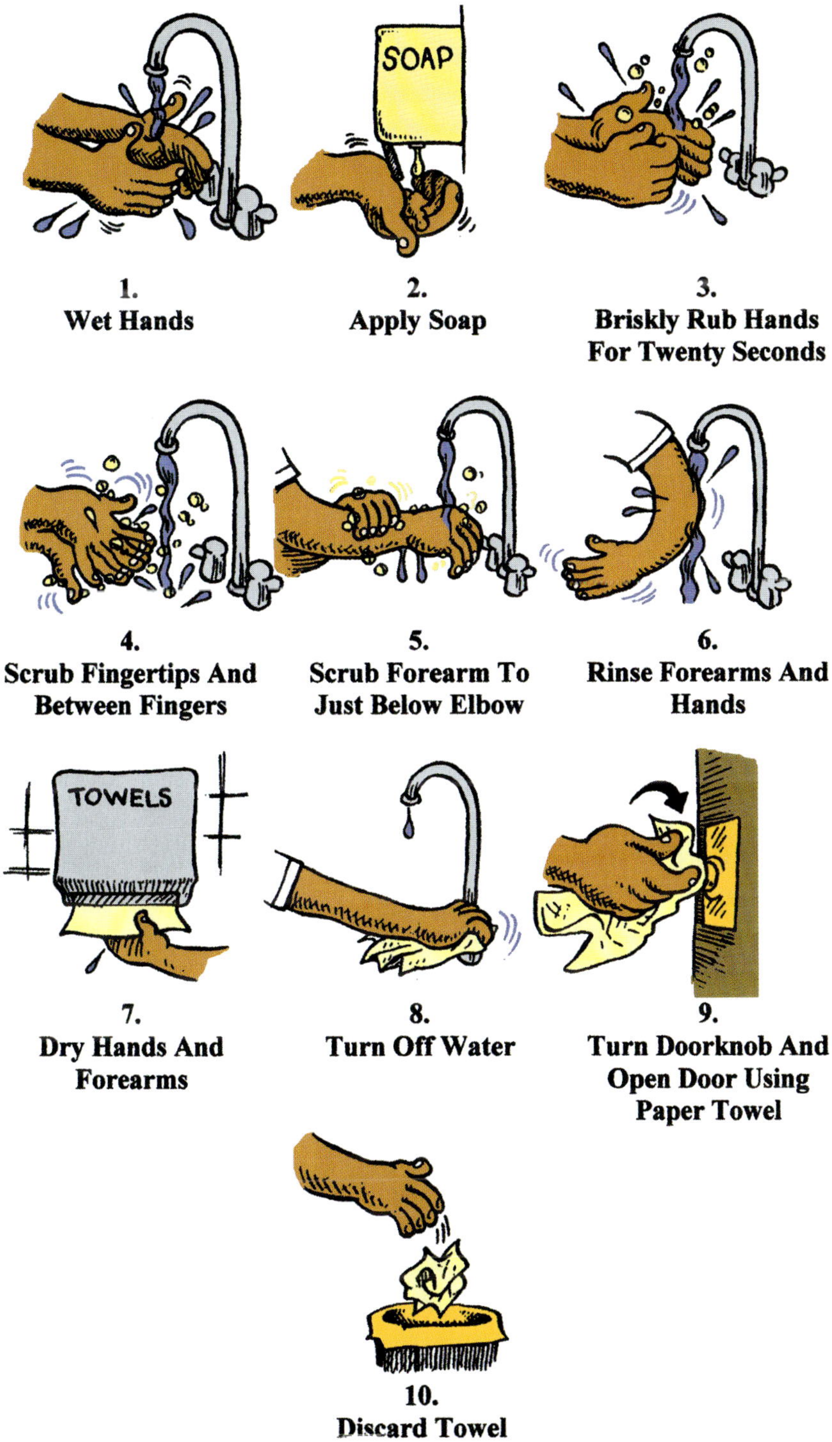

Figure 3.10 Proper Handwashing Technique

In addition to proper handwashing, fingernails should be trimmed, filed, and maintained so handwashing will effectively remove soil from under and around them. Unless wearing intact gloves in good repair, a food employee must not wear fingernail polish or artificial fingernails when working with exposed food.

According to the *FDA Food Code*, hands shall be washed in a separate sink specified as a handwashing sink (Figure 3.11). An automatic handwashing facility may be used by food employees to clean their hands. However, the system must be capable of removing the types of soils encountered in the food operation. Food employees may not clean their hands in a sink used for preparation or warewashing, or in a service sink used for the disposal of mop water and liquid waste.

Figure 3.11 Handwashing Station

Each handwashing sink must be provided with hand cleanser (soap or detergent) in a dispenser and a suitable hand-drying device.

Hand sanitizing lotions and chemical hand sanitizing solutions may be used by food employees in addition to handwashing. Proper washing helps to remove visible hand dirt and the microorganisms it contains. **Hand sanitizing lotions must never be used as a replacement for handwashing.**

After washing, the sanitizing chemical remains on the hands after application because the hands do not receive additional washing or rinsing. It is critical hand sanitizers be formulated with safe ingredients because it is likely a food employee's hands will touch food, food-contact surfaces, or equipment and utensils after using the product.

Except when washing fruits and vegetables, food employees should not contact exposed ready-to-eat foods with their hands. Instead, they should use suitable utensils such as deli tissue, spatulas, and tongs.

Retail food establishments sometimes allow their food employees to use disposable gloves to help prevent contamination of foods. Gloves can protect food from direct contact by human hands. Gloves must be impermeable, meaning they do not allow anything to penetrate the porous texture of the glove. Use of gloves is often a good idea for foods handled extensively by hands such as deli sandwiches and tacos and when conducting food demonstrations. However, it is important food handlers treat disposable gloves as a second skin. Whatever can contaminate a human hand can also contaminate a disposable glove. Therefore, whenever hands should be washed, a new pair of disposable gloves should be worn (Figure 3.12).

For example, if food handlers are wearing disposable gloves and handling raw food, they must discard those gloves, wash their hands, and put on a fresh pair of gloves before they handle ready-to-eat foods. This procedure is very important in the seafood department and other areas where raw and ready-to-eat foods are handled and displayed.

Food handlers must not handle money with gloved hands unless they immediately remove and discard the gloves. Because money is handled and exchanged by human hands, it is often contaminated with bacteria. Employees must also put on a clean pair of gloves after they complete cleaning, mopping, and similar activities within their department.

If an employee removes gloves by rolling them inside out, the inner surface of the glove is very contaminated from his or her skin. Again, when they take disposable gloves off, throw them away. Never reuse or wash disposable gloves. Always throw them away after use.

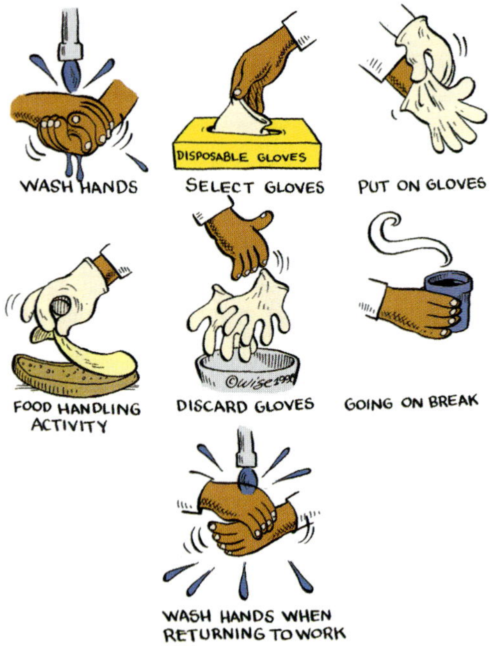

Figure 3.12 Proper Use of Disposable Gloves

Outer Clothing and Apparel

Work clothes and other apparel should always be clean. The appearance of a clean uniform is more appealing to your customers (Figure 3.13).

During food preparation and service, it is easy for a food employee's clothing to become contaminated. Whenever the outer clothing becomes contaminated, the employee should change into a clean set of work clothes.

For example, if a food handler normally wears an apron and works with raw foods, he or she should put on a clean apron before working with ready-to-eat foods. Also, it is important to teach food handlers not to dry or wipe their hands on their apron. As soon as they do that, the apron is contaminated. Smocks and aprons help to reduce transfer of microbes to exposed food. Aprons should be left in the department when employees go on break or use the restroom.

Keep in mind, however, protective apparel is similar to a disposable glove. They no longer protect food when contaminated.

Hats, hair coverings or nets, and beard restraints discourage employees from touching their hair or beard. These restraints also prevent hair from falling into food or onto food-contact surfaces. Most state and local jurisdictions require food employees to effectively keep their hair from contacting exposed food, clean equipment, utensils and linens, and unwrapped single-use articles.

Figure 3.13 Employees Must Wear Proper Attire during Food Preparation and Service

Personal Habits

Personal hygiene means good health habits including bathing, washing hair, wearing clean clothing, and frequent hand washing. Poor personal habits are serious hazards in retail food establishments. A food employee's fingers may be contaminated with saliva during eating and smoking. Saliva, sweat, and other body fluids can be harmful sources of contamination if they get into food.

Supervisors should enforce rules against eating, chewing gum, and smoking in food preparation, service, and warewashing areas. The *FDA Food Code* permits food employees to drink beverages to prevent dehydration. The beverage must be in a covered container. The container must be handled in a way that prevents contamination of the employee's hands, the container, exposed food, equipment, and single-use articles.

Jewelry, including medical information jewelry on hands and arms, has no place in food production and warewashing areas. Rings, bracelets, necklaces, earrings, watches, and other body part ornaments can harbor germs that can cause foodborne illness. Jewelry can also fall into food causing a physical hazard. A plain wedding band, that does not contain a stone, is the only piece of jewelry that may be worn in food production and warewashing areas.

Personal Health

In an attempt to reduce the risk caused by sick food employees (Figure 3.14), **the *FDA Food Code* requires employees to report to the person in charge when they have been diagnosed with:**

- *Salmonella* Typhi
- *Shigella spp.*
- Shiga toxin-producing *Escherichia coli*
- Hepatitis A.

Food employees must also report when they have symptoms of intestinal illness or flu-like symptoms (such as vomiting, diarrhea, fever, sore throat, or jaundice) or a lesion containing pus such as a boil or infected wound that is open or draining.

If a food employee is directly or indirectly exposed to *Salmonella* Typhi, *Shigella spp.*, Shiga toxin-producing *Escherichia coli*, or Hepatitis A, it must be reported to the supervisor. All these agents are easily transferred to foods and are considered severe health hazards. The person in charge shall notify the regulatory authority when a food employee is diagnosed with an illness due to *Salmonella* Typhi, *Shigella spp.*, Shiga toxin-producing *Escherichia coli*, or Hepatitis A virus.

If a food employee has been exposed to any of these agents, he or she may be excluded from work or be assigned to restricted activities that do not involve contact with food. Employees diagnosed with one of these diseases must not handle food or have contact with clean equipment, utensils, linens, or unwrapped single-service utensils.

Figure 3.14 Sick Employees Must Not Work With Food

Staphylococcus aureus bacteria are often found in infected wounds, cuts, and pimples. Infected wounds should be cleaned and completely covered by a dry, tight-fitting, impermeable bandage. Cuts or burns on a food employee's hands must be thoroughly cleaned and bandaged and covered with a clean disposable glove.

When a foodborne illness outbreak occurs and an employee is suspected of carrying the disease, he or she must not work until certified as safe by a licensed physician, nurse practitioner, or physician assistant. Exclusions and restrictions for those exposed to or experiencing symptoms of a foodborne illness are specified in Appendix C of this book. **The health and hygiene of food employees are extremely important in safe food management.**

To date, there has not been a medically documented case of Acquired Immune Deficiency Syndrome (AIDS) transmitted by food. There is no evidence the virus can be transferred to foods after human contact. Therefore, AIDS is not considered a foodborne illness. The Americans with Disabilities Act (ADA) prohibits discrimination against people with disabilities in jobs and public accommodations. Employers may not fire or transfer individuals who have AIDS or test positive for the HIV virus away from food-handling activities. Employers must also maintain the confidentiality of employees who have AIDS or any other illness.

Cross Contamination

Contaminated food contains germs or harmful substances that can cause foodborne illness. The transfer of germs from one food item to another is called **cross contamination**. This commonly happens when germs from raw food are transferred to a cooked or ready-to-eat food via contaminated hands, equipment, or utensils.

For example, bacteria from fish can be transferred to ready-to-eat seafood (such as cooked shrimp and imitation crab) in a display case where the products are not properly separated or by an employee's gloves that are not changed between handling raw and ready-to-eat foods. Cutting boards are also common vehicles for cross contamination. The leftover juices from raw chicken can lead to contamination of ready-to-eat salad items as illustrated in Figure 3.15.

Figure 3.15 Cross Contamination of a Cutting Board

Cross contamination also happens when raw foods are stored above ready-to-eat foods. Juices from the raw product can drip or splash onto a ready-to-eat food as in Figure 3.16. This poses a serious health risk because the ready-to-eat salad items will not be cooked to destroy the microorganisms prior to being eaten.

Avoiding Cross Contamination

There are many things you can do to avoid, or at least reduce, opportunities for cross contamination during food preparation and service. An important part of food safety is to impress upon your employees the importance of clean hands and clothing. Stress the importance of when and how to properly wash hands. Provide ongoing supervision to ensure employees remain clean and apply safe food-handling practices when working with food.

To avoid cross contamination by way of food-contact surfaces (i.e., cutting boards, knife blades, slicers, and sandwich preparation areas), an effective in-house cleaning and sanitation program is important. You will learn more on this subject in Chapter 6, *Cleaning and Sanitizing Operations* on page 215.

Figure 3.16 Never Put Raw Food Over Cooked or Ready-to-Eat Food

If a food-contact surface is not properly cleaned and sanitized, bacteria can accumulate and grow on the surface. Once bacteria get on the surface, they can be difficult to remove. To avoid such buildup, always clean and sanitize food-contact surfaces thoroughly before working with ready-to-eat foods and foods that will not be cooked.

The following preventive measures can be used to eliminate the possibility of cross contamination between products:

- Use separate equipment, such as cutting boards, for raw foods and ready-to-eat foods (color coding may be helpful for this task)
- Use clean and sanitized equipment and utensils for food production
- Avoid bare hand contact with ready-to-eat food
- Prepare ready-to-eat foods first–then raw foods
- Prepare raw and ready-to-eat foods in separate areas of the food establishment
- Store cooked and ready-to-eat foods over raw products
- Keep raw and ready-to-eat foods separate during storage
- Keep species of meat and poultry separate in meat departments.

Always keep raw foods separate from ready-to-eat foods. In the refrigerator, ready-to-eat foods must be stored above raw foods. Display cases, such as those used to display seafood items, should be designed to keep raw and cooked food items separate. In addition, separate buckets for in-place sanitizing solutions and wiping cloths should be used for cleaning food-contact surfaces in raw and ready-to-eat food production areas.

Other Sources of Contamination

Certain raw foods, such as fruits and vegetables, should be treated like ready-to-eat foods. Always wash these foods before use. Washing removes soil and other contaminants. Chemicals may be used to wash raw whole fruits and vegetables. These chemicals must be nontoxic and meet the requirements set forth in the Code of Federal Regulations under title 21CFR173.315.

Utensils used to dispense and serve foods can also be a source of food contamination. Utensils should be properly labeled to identify the type of food they are used to dispense. During hot- or cold-holding of foods, the utensil should be stored in the food. This helps to prevent contamination from employees or customers in self-service areas. It also keeps the utensil that contains food, out of the temperature danger zone. The dispensing utensils (scoops) for ice and dry bulk food should be clean and kept in an area that is protected from contamination. Scoops and tongs used in customer service areas also need to be labeled and kept clean.

Customer self-service areas for bakery items and candy are becoming more popular. Contamination by food allergens, such as peanuts, has occurred in both of these areas. Be sure the scoop used for peanut butter cups is not used for hard candies. To ensure dispensing utensils are used for the intended foods, many retail food establishments will attach the dispensing utensil to the display case with a cable or chain.

Animals are not allowed in food establishments unless they are being used for support or special service (i.e., guide dogs for the blind). It is very important food handlers do not touch animals during food preparation and service. If employees should touch an animal for any reason, they must wash their hands before returning to work.

Germs from a employee's mouth can be transferred to food when the employee uses improper tasting techniques. A food employee may not use a utensil more than once to taste food that is to be sold or served.

Animals and pests are common sources for food contamination (Figure 3.17). Rodents and pests usually enter food establishments during delivery or when garbage facilities are not properly maintained. A good integrated pest management (IPM) program should be established and maintained in any food establishment. You will learn about IPM programs in Chapter 7, *Environmental Sanitation and Maintenance.*

Figure 3.17 Other Sources of Contamination

Make Sure the Work Area is Clean and Sanitary

Anything that comes in contact with food must be clean and sanitary. This includes human hands, equipment, utensils, storage and holding areas, and self-service areas for customers. In order to protect food from contamination, effective cleaning and sanitizing procedures must be implemented and monitored. The goal of cleaning is to remove visible soil. The goal of sanitizing is to reduce the number of harmful microbes that may be present on a clean surface. Chapter 6 of this book is dedicated to describing different elements of cleaning and sanitizing programs. While cleaning and sanitizing are important in all areas of a retail food establishment, they are especially critical in areas where ready-to-eat foods are handled and displayed.

Summary

In the case study presented at the beginning of this chapter, the employees who tested positive for Hepatitis A were treated and ordered to stay home until their physician authorized them to return to work. Employees in the seafood department who did not test positive for the virus were given Hepatitis A vaccinations. Training sessions on proper handwashing were provided for all employees in the department. The training sessions taught employees when and how to wash their hands and stressed the importance of avoiding bare hand contact with food by wearing disposable gloves and using utensils.

Temperature abuse can occur during receiving, storing, cooking, cooling, reheating, hot-holding, cold-holding, and display of foods. Depending on the food type, there are specific temperature requirements to ensure food safety. These requirements are discussed in the next chapter. A good rule to follow in any retail food establishment is, "Keep It Clean, Keep It Hot, Keep It Cold, Or Don't Keep It!"

Food employees should always use good health and hygiene practices. Clean clothing, hair restraints, and proper handwashing practices are fundamental in safe food management. Effective supervision and enforcement of proper procedures form the foundation of a successful food operation.

Control cross contamination with proper cleaning and sanitizing. Avoid cross contamination from one food to another. Keep foods separate and store raw foods below cooked and ready-to-eat foods. Always use proper food-handling techniques.

Keep foods at proper temperature, use good personal hygiene, control cross contamination, and implement good cleaning and sanitizing programs. These are the essentials for preventing foodborne illness.

Discussion Questions (Short Answer)

1. Name the four elements of foodborne illness prevention identified in this chapter.
2. When purchasing and using a food thermometer, what are the six important things you need to remember?
3. How frequently should thermometers be calibrated? Describe the two methods for calibrating food thermometers.
4. Under what conditions are foods "time and temperature abused"?
5. What is meant by poor personal hygiene and how can this lead to foodborne illness?
6. What does the term cross contamination mean?
7. Why is cleaning and sanitizing important?

Quiz 3 (Multiple Choice)

1. The No. 1 contributing factor leading to foodborne illness in retail food establishments is:
   a. Improper holding temperature of foods.
   b. Cross contamination.
   c. Poor personal hygiene.
   d. Inadequate cleaning and sanitizing programs.
2. Foodborne illness can be caused by:
   a. Poor personal hygiene.
   b. Cross contamination.
   c. Temperature abuse.
   d. All of the above.

3. Regarding food thermometers, which statement is false?
   a. They should be calibrated.
   b. They should measure temperatures between 41° and 140°F (5° and 60°C).
   c. They should measure temperatures between 0°and 220°F (-18° and 104°C).
   d. They should be approved for use in foods.
4. Good personal hygiene includes:
   a. Using hand sanitizers instead of handwashing.
   b. Keeping hands and clothes clean and sanitary.
   c. Wearing attractive uniforms.
   d. Cleaning and sanitizing food-contact surfaces.
5. Food employees should wash their hands after which of the following?
   a. Taking out the trash.
   b. Touching their face.
   c. Handling raw food.
   d. All of the above.
6. Cross contamination is a term used to describe the transfer of a foodborne hazard from one food to another:
   a. By a food employee's hands.
   b. From a cutting board.
   c. From a knife blade.
   d. All of the above.
7. A good way to prevent cross contamination of foods is to:
   a. Keep raw and cooked foods separate.
   b. Properly clean and sanitize food-contact surfaces.
   c. Properly wash hands.
   d. All of the above.
8. Which of the following is an accepted personal hygiene practice?
   a. Wearing jewelry and false fingernails.
   b. Smoking and eating in food production areas.
   c. Wearing caps and hats.
   d. Wiping hands on a soiled apron.

9. After proper cooking, all foods that are to be held cold must be:
   a. Cooled quickly and held at 41°F (5°C) or below.
   b. Cooled quickly and held at 70°F (21°C) or above.
   c. Stored at room temperature until served.
   d. Cooled slowly and held at 50°F (10°C) or below.
10. After proper cooking, all foods that are to be held hot must be held at:
   a. 165°F (74°C) or above.
   b. 140°F (60°C) or above.
   c. Room temperature until served.
   d. 120°F (49°C) or above.

Answers to the discussion and multiple choice questions are provided in Appendix A.

References/Suggested Readings

Centers for Disease Control and Prevention (2000). *Surveillance for Foodborne Disease Outbreaks—United States, 1993-1997.* March 17, 2000. Atlanta, GA; U.S. Department of Health and Human Services.

Council for Agricultural Sciences and Technology (1995). Prevention of Foodborne Illness. Dairy. *Food and Environmental Sanitation.* Vol. 15(6), 357-367.

Food and Drug Administration (2001). *2001 Food Code.* U.S. Public Health Service. Washington, DC.

Food Safety and Inspection Service (1996). *Nationwide Broiler Chicken Microbiologic Baseline Data Collection Program, 1994-1995.* U.S. Department of Agriculture. Washington, DC.

Suggested Web Sites

Gateway to Government Food Safety Information
www.foodsafety.gov

United States Department of Agriculture (USDA)
www.usda.gov

Centers for Disease Control and Prevention (CDC)
www. cdc.gov

United States Food and Drug Administration (FDA)
www.fda.gov

USDA/FDA Food and Nutrition Information Center
www.nal.usda.gov/fnic/

Partnership for Food Safety Education
www.fightbac.org

Food Safety Day
www.foodsci.purdue.edu/publications/foodsafetyday

4 FOLLOWING THE FOOD PRODUCT FLOW

Salmonella Outbreak Associated with Cross Contamination of Food

Twenty-four customers of a small retail food establishment became ill with a salmonella infection after eating ham and cheese sandwiches prepared at the establishment. The victims experienced a variety of symptoms including headache, nausea, vomiting, fever, diarrhea and abdominal pain.

Inspection of the premises indicated the counter top and work surface in the delicatessen were not cleaned and sanitized after raw chicken was prepared for frying earlier in the morning. The raw chicken and then the ready-to-eat breakfast sandwiches were prepared on the same counter and bacteria from the raw chicken were transferred to the ready-to-eat breakfast sandwiches.

What measures could have been taken to prevent this foodborne outbreak?

Learning Objectives

After reading this chapter, you should be able to:

- ▲ Recognize codes and symbols used to designate food products that have been inspected by governmental agencies.
- ▲ Apply purchasing and receiving procedures that enhance the protection of food products.
- ▲ Evaluate equipment used to transport food products to retail food establishments.
- ▲ Use approved devices to measure temperatures in food products safely and accurately.
- ▲ Recognize product defects and refuse acceptance of products that do not meet established food safety criteria.

- ▲ Identify product temperatures required at receiving and storage.
- ▲ Discuss safe methods to thaw frozen foods.
- ▲ Identify internal temperature requirements for cooking foods.
- ▲ Explain the proper methods used to cool foods.
- ▲ Discuss the importance of employee health and hygiene related to food flow.
- ▲ Employ measures to prevent contamination and cross contamination of foods.

Essential Terms

Additives	**Inspection for wholesomeness**
Aseptic packaging	**Irradiation**
Cold-holding	**Leftovers**
Cooking	**Pasteurization**
Cooling	**Reduced Oxygen Packaging (ROP)**
First In, First Out (FIFO)	**Reheating**
Grade standards	**Sensory evaluation**
Hermetic packaging	**Sous vide**
Hot-holding	**Wholesome**

Buying from Approved Sources

Food quality and safety in retail food establishments begins with buying foods and ingredients from approved sources. Approved sources are processors and suppliers that comply with applicable federal, state, and local food safety laws and regulations. Approved sources are routinely inspected by regulatory inspectors to make sure they follow good manufacturing practices. Regulatory agencies work closely with each other and the food industry to assure the safety and **wholesomeness** of foods produced and sold for human consumption. A food is wholesome if it is safe to eat or promotes the health of customers. A food establishment or retail company should develop food safety buyer specifications to assure they are receiving the safest and highest quality possible.

Foods prepared in a private home are not considered to be from an approved source. Therefore, they must not be used or offered for sale in retail food establishments. Aunt Betty may make the best cookies in town, but you cannot sell her cookies at your establishment unless the health department routinely inspects and approves the kitchen where her cookies are baked. The use of "home-canned" food is also prohibited because of the high risk of foodborne illness, especially botulism.

Inspecting Delivery Vehicles

Suppliers must deliver food products to your retail food establishment in vehicles that are clean and in good repair. Delivery vehicles must also:

- Maintain perishable and potentially hazardous foods at safe temperatures during transport
- Be loaded in a manner that separates food items from non-food items (detergents, household cleaners, and pesticides) to prevent contamination and cross contamination
- Protect food packages from becoming damaged and torn during transit.

When delivery vehicles arrive at your establishment, receiving personnel should inspect them. Specific items to look for during these inspections are:

- Cleanliness of the cargo area
- Temperature of refrigerated and frozen storage areas (if applicable)
- Proper separation of food and non-food items
- Signs of insect, rodent, or bird infestations
- Damaged packages that might result in contamination of food items.

Strategies for Determining Food Quality

Sensory evaluation is a commonly used method for making routine quality determinations on foods received at retail food establishments.

This type evaluation involves using the senses of smell, touch, sight, and sometimes taste (Figure 4.1).

Figure 4.1 Inspect Foods When They Are Received

As a first step, foods should be observed for color, texture, and visual evidence of spoilage. Quite often, spoilage is easily seen as slime formation, mold growth, and discoloration. Product containers should also be checked for tears, punctures, dents, or other signs of damage.

Flavor is a combination of smell and taste. Spoiled foods frequently give off foul odors indicative of compounds such as ammonia and hydrogen sulfide (the smell of rotten eggs). These odors, caused by the breakdown of proteins through bacterial action, are usually very easy to smell.

The flavor of spoiled foods can range from loss of good characteristic taste to the development of objectionable tastes. Spoilage due to yeasts produces bubbles and an alcoholic flavor or smell. Milk develops an acidic taste and is often bitter when it spoils.

The feel of spoiled foods is frequently determined by the type of food and the spoilage organism involved. Some foods feel slimy, whereas others may feel mushy.

The quality and safety of a food are affected by many factors. A food that shows no signs of spoilage may not always be safe. Spoilage cannot be used as the only indicator of food safety.

Measuring Temperatures at Receiving and Storage

Maintaining safe product temperature is a critical part of your food safety system. According to foodborne disease investigations, improper temperature control during food preparation and service is one of the leading contributors to foodborne illness. Temperature-measuring devices are used in retail food establishments to measure temperatures of food, water, and the air of food storage areas (refrigerators, ovens, etc.).

You can measure the approximate temperature of packaged foods without opening the package. Place the probe of the thermometer or thermocouple between two packages of food or fold the package around the stem or probe to make good contact with the packaging. Packaged food temperatures may also be approximated with the use of infrared thermometers. The photographs in Chapter 3, Figure 3.6 on page 86, illustrate different ways to measure the temperature of packaged and prepared foods.

To measure the temperature of an unpackaged food, insert the sensing element of the thermometer or thermocouple into the thickest part of the food. To avoid cross contamination of food items, be sure to clean, sanitize, and air dry the probe of the thermometer before going from a raw animal food to a cooked or ready-to-eat food.

Cold- or hot-holding equipment used for storing potentially hazardous foods should be equipped with an indicating or recording thermometer to measure the temperature of the storage environment. Equipment thermometers are either built into a piece of equipment or are fastened onto shelving or other apparatus. Position them where they can be easily read. Place the sensor portion of the thermometer in the warmest part of a refrigeration unit or in the coolest part of a hot food storage unit. You will have to use a portable thermometer or thermocouple to measure air temperature in the storage compartment if the equipment does not include an indicating or recording thermometer.

Following the Flow of Food

The flow of food at a retail food establishment begins with receiving and storage. As foods are delivered, they are inspected by receiving personnel and then placed quickly into storage. Two or more departments in the establishment may share some storage areas. Others will be used by a single department such as the deli, dairy, meat, or seafood department.

From storage, foods and ingredients flow into the preparation and handling stages of production or directly into a display case. Preparation and handling commonly involve multiple steps or processes. Some common preparation steps in the meat department would be grinding, cutting, trimming, packaging, and cold-holding during display. Some examples of production activities in a deli would include slicing luncheon meats and cheeses, dispensing ready-to-eat salads from bulk containers into display trays and from display trays into individual serving containers, making sandwiches, cooking chicken, and hot- and cold-holding foods during display. A simple diagram showing the overall flow of food in a retail food establishment is presented in Figure 4.2. A more detailed explanation of the steps in the flow of food will be presented below.

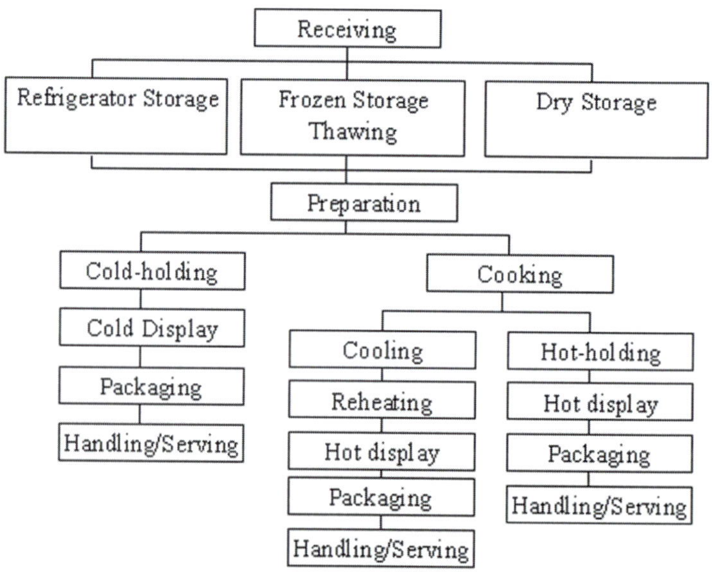

Figure 4.2 Food Flow Diagram

Receiving

Employees responsible for receiving products must carefully inspect all incoming food supplies to make sure they are at proper temperature, in sound condition, and free from filth or spoilage.

As a first step, receiving personnel should observe foods for color, texture, and visual evidence of spoilage. Quite often, spoilage is easily visible as in the case of slime formation, mold growth, and discoloration. Associates should also check product containers for tears, punctures, dents, or other signs of damage.

Proper receiving requires a knowledgeable person who follows specific guidelines. Poor receiving procedures increase the chance of:

- ▲ Theft
- ▲ Acceptance of underweight merchandise
- ▲ Contamination
- ▲ Shrink
- ▲ Acceptance of products that do not meet specifications.

Anticipate the arrival of deliveries and be sure to have enough staff and space on hand to receive products quickly and correctly. Schedule deliveries during off peak periods of the day. Employees must receive and store incoming shipments as soon as they arrive. Always make certain food products are delivered in clean, well-maintained vehicles that protect the food from contamination and maintain proper product temperature.

Distressed merchandise that is damaged, spoiled, or otherwise unfit for sale or use in a retail food establishment must be properly disposed of, held by the facility for credit, or returned to the distributor. It is important distressed products not be mistaken for safe and wholesome products and cause contamination of other foods, equipment, utensils, linens, or single-service or single-use articles. To prevent this from happening, these products must be segregated and held in designated areas that will keep them separate from food, equipment, utensils, linens, and single-service and single-use articles.

Whatever the size of the establishment, receiving requires:

- ▲ Prompt handling
- ▲ Exacting quality control procedures
- ▲ Trained staff who use good judgment and experience in interpreting:
    - ▼ Product specifications
    - ▼ Coding
    - ▼ Temperature measurement
    - ▼ Proper handling of distressed merchandise.

Packaged Foods

The foods in a retail food establishment come in many different types of packages. Examples of food packaging include cans, bottles, jars, pouches, tubs, trays, bags and boxes. The common purpose of the package is to:

- ▲ Protect the contents from contamination
- ▲ Provide a source of information about its nutritional contents
- ▲ Provide advertising material
- ▲ Make the product more convenient for customers to transport, prepare, and serve.

Dry foods such as flour, sugar, rice, and beans are commonly packaged in bags. These are not potentially hazardous foods and can be safely stored at room temperature. At receiving, packages of dry food must be inspected for tears and other signs of damage and contamination by chemicals and other substances that could cause foodborne illness.

Reduced Oxygen Packaging (ROP)

Oxygen in the air increases both the chemical breakdown and microbial spoilage of many foods. Many food processors use **reduced oxygen packaging** to help overcome the effects of oxygen and preserve foods. Some examples of reduced oxygen packaging include vacuum packaging, modified atmosphere packaging, and sous vide.

Vacuum packaging removes air from a package and hermetically seals the package so a near-perfect vacuum remains inside. The term **hermetic** refers to a container completely sealed to prevent the entry and loss of gases and vapors. The most commonly used hermetic packages are metal cans and glass jars. These containers will also stop the entry of bacteria, yeasts, molds and other types of contamination as long as they remain undamaged.

Upon receiving, metal cans must be checked for defects such as:

- ▲ Leaks
- ▲ Bulging ends
- ▲ Dents
- ▲ Broken seals
- ▲ Damage along seams
- ▲ Penetrating rust

▲ Missing labels.

Do not accept cans if they leak or bulge at either end. Swollen ends on a can indicate gas is being produced inside. This gas may be caused by a chemical reaction between the food and the metal in the container, or it may be caused by bacteria and other microbes inside the can.

Dents in cans do not harm the contents unless they have actually penetrated the can or the seam. Dents found in the side seams or end seams of a can are the most important. Damage to these areas can affect the physical integrity of the can and may allow microorganisms to enter through tiny pinhole leaks. Shipments with many dented cans or torn labels indicate poor handling and storage procedures by the supplier. Figure 4.3 shows some common defects found in cans.

Figure 4.3 Normal and Defective Cans

(Courtesy Damage Recovery Systems, Inc.)

Modified atmosphere packaging (MAP) is used with a wide range of products including meat, fish, pre-cut lettuce, baked products, cheese, coffee, nuts, and dried fruit (Figure 4.4). The MAP process has been successful in extending the shelf life of many products, and it reduces the amount of additives and preservatives required to prevent deterioration of the food. However, since low oxygen environments can set up conditions conducive to the growth of anaerobic bacteria like *Clostridium botulinum* (See Chapter 2 on page 46), proper handling of MAP foods is essential. MAP technology requires adequate refrigeration be maintained during the entire shelf life of potentially hazardous foods. During the receiving process, employees must inspect potentially hazardous foods in modified atmosphere packages to make sure the package is in sound condition and the food is at 41°F (5°C) or below when it arrives at the establishment.

Sous vide is a French term for "without air." It is a specialized packaging process where fresh raw foods are sealed in plastic pouches and the air is removed by vacuum. The pouches are then cooked at a low temperature and rapidly cooled to 38°F (3°C) or below or frozen. The low cooking temperature of the sous vide process kills spoilage microorganisms. However, it does not destroy *C. botulinum* bacteria and their spores. Therefore, potentially hazardous foods processed using sous vide technology must be kept out of the food temperature danger zone during transport, storage, and display. Receiving personnel must inspect sous vide products to make sure the package is in sound condition and the food is at 41°F (5°C) or below or solidly frozen as it arrives at the establishment. If extended shelf life is desired, a temperature of 38°F (3°C) or lower must be maintained at all times. After receiving, the products should be moved quickly into refrigerated or frozen storage until ready for use.

a. Pre-cut salad items

b. Vacuum packaged meat

Figure 4.4 Examples of MAP Foods

Food **irradiation** is a preservation technique used by some food processing industries. This process involves exposing food to certain forms of radiation in order to destroy disease-causing microorganisms and delay spoilage. The FDA has approved food irradiation for a variety of foods including fruits, vegetables, grains, spices, poultry, pork, lamb, and, more recently, ground beef.

The acceptance of irradiated foods has been limited due to customers' concerns about the safety of foods preserved in this manner. Contrary to many myths, irradiated food is not radioactive and does not pose a risk to the health and safety of people who eat it. Foods processed with irradiation are just as nutritious and flavorful as other foods that have been cooked, canned, or frozen.

Federal law requires irradiated food must be labeled with the international symbol for irradiation called a "radura" (Figure 4.5). This symbol must be accompanied by the words "Treated with Irradiation" or "Treated with Radiation."

Irradiation of food can effectively reduce or eliminate pathogens and spoilage microbes while maintaining the quality of most foods. This is a technology proven safe and should be welcomed by customers as an effective food preservation technique.

Figure 4.5 Radura

Red Meat Products

Most red meat and meat products sold in the United States come from cattle (beef), calves (veal), hogs (ham, pork, and bacon), sheep (mutton), and young sheep (lamb). These products are inspected for wholesomeness by officials of the U.S. Department of Agriculture (USDA) or state agencies. Animals must be inspected for wholesomeness to make certain they are free of disease and unacceptable defects. The USDA also offers voluntary meat grading services. Grades for meat represent the culinary quality or palatability of the meat and are not measures of product safety. Figure 4.6 contains examples of inspections and grading stamps applied to products approved for use.

Inspection Grade

Figure 4.6 USDA Inspection and Grade Stamps for Beef, Veal, and Lamb

Meat and meat products in retail food establishments are available in several forms such as fresh, frozen, cured, smoked, dried, and canned. Since raw meats are potentially hazardous foods, never accept them if there is any sign of contamination, temperature abuse, or spoilage.

Never accept fresh meat if the product temperature exceeds 41°F (5°C) at delivery. Fresh meat should be firm and elastic to the touch. Loss of texture in meat and accumulation of liquid in the package indicate the onset of bacterial spoilage. The color should be bright cherry red and a purple/red color if vacuum packed or as a ROP food. The visible fat will also change from white to yellow as meat undergoes spoilage. Meat that has begun to turn brown or green indicates spoilage. Off odors are frequently indicators of spoilage. Meat that has an offensive odor or "rotten egg" smell should not be offered for sale. Sliminess is another characteristic of spoilage and is caused by bacterial growth on the surface of meat. Control of factors that cause spoilage and sliminess also extends the shelf life of meat products.

Frozen meats should be solidly frozen when they arrive at your establishment. Look for signs of freezing and thawing and refreezing such as frozen blood juices in the bottom of the container or the presence of large ice crystals on the surface of the product. Frozen meats should be packaged to prevent freezer burn.

Fresh meat should be moved into refrigerated storage as quickly as possible. For fresh meat, evaluate the date on which the product was packed. A company policy should be established on how long beyond the date of pack the product can be accepted and sold. Frozen products should be moved from the delivery truck to the freezer while they are still solidly frozen.

Game Animals

Game animals are not allowed for sale in retail food establishments unless they meet federal code regulations. This ban does not apply to commercially raised game animals approved by regulatory agencies, field dressed game allowed by state codes, or exotic species of animals that must meet the same standards as those of other game animals.

Game animals commercially raised for food must be raised, slaughtered, and processed according to standards used for meat and poultry. Common examples of animals raised away from the wild and used for food are farm-raised buffalo, ostrich, and alligator. The USDA inspects the slaughter and processing of this meat in the usual manner.

Some states permit game animals, such as deer, bear, and elk that have been killed and dressed in the field, to be sold in retail food establishments. Game meat must be dressed soon after the kill to prevent rapid growth of bacteria already present in the meat. Next, the meat must be chilled rapidly, transported in a sanitary manner, and processed in an approved facility. Veterinarians are frequently appointed to inspect meat for contamination that would harm humans.

Poultry

All poultry products must be inspected for wholesomeness by USDA or state inspectors. Inspection for wholesomeness involves an examination of the poultry to make certain it is wholesome and not adulterated. Adulterated food contains filth or is otherwise decomposed and unfit for human consumption. The poultry is examined live before slaughter, during evisceration (removal of the internal organs), and during or after packaging. Inspected poultry products carry a USDA seal for wholesomeness on the individual package or on bulk cartons. Examples of inspection and grade stamps for poultry are shown in Figure 4.7.

Poultry is also graded for quality. Top quality birds receive a Grade A rating. To earn a Grade A rating, the poultry must have good overall shape and appearance, be meaty, practically free from defects, and have a well-developed layer of fat in the skin.

Inspection **Grade**

Figure 4.7 USDA Inspection and Grade Stamps for Poultry

Poultry products support the growth of disease-causing and spoilage microorganisms. The intestinal tract and skin of poultry may contain a variety of foodborne disease bacteria, including *Salmonella spp.* and *Campylobacter spp.* The near neutral pH, high moisture, and high protein content of poultry make it an ideal material for bacteria to grow in and on (Figure 4.8).

Figure 4.8 Spoiled Poultry

Poultry products are also vulnerable to spoilage caused by enzymes and spoilage bacteria. Most spoilage bacteria come from the live bird's skin or intestinal tract. The bacteria grow on the skin of slaughtered birds and inside the birds' carcasses. Spoilage is indicated by meat tissue that is soft, slimy, and has an objectionable odor. Stickiness under the wings is another sign of poultry spoilage caused by drying or exposure. Reject poultry that is discolored or has darkened wing tips or sticky skin. At the time of delivery, poultry should be packaged on a bed of ice that drains away from the meat as it melts and held at or below 41°F (5°C).

Eggs

Most retail food establishments sell and use eggs in one form or another. Eggs are usually purchased by federal grades, the most common being AA, A, and B (Figure 4.9). Grades for eggs are based on exterior and interior conditions of the egg. Eggs cannot receive a grade when the shell is dirty, cracked, or broken.

Figure 4.9 USDA Inspection and Grade Stamps for Eggs

The USDA reports approximately 50 billion eggs are sold in the United States each year. *Salmonella enteritidis* bacteria are present in about 1% (approximately 500 million or one in 20,000) of the eggs sold. The bacteria enter the yolk of the egg as it is formed inside the hen. The eggshell surface may contain *Salmonella spp.* bacteria, especially if the shell is soiled with chicken droppings. Even if the shell is not cracked, bacteria can enter through the pores in the egg's shell. Raw shell eggs should be clean, fresh, free of cracks or checks, and refrigerated at an ambient air temperature of 45°F (7°C) or below when delivered. Shell eggs that have not been treated to destroy all viable *Salmonella* shall be stored and displayed in refrigerated equipment that maintains an ambient temperature of 45°F (7°C) or less. These eggs must be labeled to include safe handling instructions. The egg, when opened, should have no noticeable odor, the yolk firm, and the white should cling to the yolk. Reject eggs that are dirty or cracked, and remember washing eggs only increases the possibility of contamination.

An egg product is defined as an egg without its shell. As a safeguard against *Salmonella spp.*, the FDA requires all egg products, such as liquid, frozen, and dry eggs, be pasteurized to render them *Salmonella*-free. Pasteurized egg products should be in a sealed container and kept at 41°F (5°C). Egg containers should carry labels that verify the contents have been pasteurized. These products are well suited for facilities that offer food to people in highly susceptible populations.

Fluid Milk and Milk Products

This food group includes milk, cheese, butter, ice cream, and other types of milk products. When receiving milk and milk products, make certain they have been pasteurized. Pasteurization destroys all disease-causing microorganisms in the milk and reduces the total number of bacteria, thus increasing shelf life. All market milk must be Grade A quality. Pasteurization also destroys natural milk enzymes that might shorten the shelf life of the products.

Milk that is marked "UHT" pasteurized has been heated to ultra-high temperatures and placed in **aseptic packaging**. UHT products can be stored safely for several weeks if kept under refrigeration. These products can be stored without refrigeration for short periods of time (Longreè and Armbruster). Individual creamers are sometimes processed in this manner.

Fluid Milk

Under the Pasteurized Milk Ordinance, fluid milk can be received at 45°F (7°C) or less. Once received, it should be refrigerated immediately upon delivery and held at 41°F (5°C) or below. Individual containers of milk should be clearly marked with an expiration date and the name of the dairy plant that produced it. Check the expiration date of all dairy products before using them.

Cheese

Cheese should be received at 41°F (5°C) and checked for the proper color, flavor, and characteristics. Reject the product if it contains mold that is not a normal part of the cheese or if the rind or package is damaged.

Butter

Butter is made from pasteurized cream. Since disease-causing and spoilage bacteria and mold may grow in butter, handle the product as a perishable item. The most common type of deterioration in butter is the development of a strong rancid odor and flavor. Check to see butter has a firm texture, even color, and is free of mold. Package butter should be received at 41°F (5°C), intact and provide protection for the contents.

Staphylococcus aureus toxin has been the cause of foodborne disease outbreaks where butter and cheese were involved. *Staphylococcus aureus* bacteria do not require as much moisture as other pathogens. They can thrive in salty foods that would prevent the growth of other types of pathogens. Therefore, cheese and butter must be handled carefully and kept out of the temperature danger zone except as required during production.

Fish

Fish includes finfish harvested from saltwater and freshwater, and seafood that comes mainly from saltwater. Catfish and trout are two of the most popular types of freshwater finfish. Tuna, snapper, and flounder are saltwater finfish that are very popular with American customers. Seafood consists of molluscan shellfish and crustaceans. Molluscan shellfish include oysters, clams, mussels, and scallops. Crustaceans include shrimp, lobster, and crab. Oysters, shrimp, catfish, salmon, and a few other types of finfish and seafood are being raised on fish farms using a technique called aquaculture.

Fish should be received at 41°F (5°C) or below and shellfish may be received at 45°F (7°C) or below. For better quality and shelf life, the optimum fish and shellfish receiving temperature often ranges from 30°F (-1°C) to 34°F (1°C). These products are generally more perishable than red meats, even when stored in a refrigerator or freezer. They are commonly packed in self-draining ice to prevent drying and to maximize the shelf life of the food. The slime covering the outside of fish contains a variety of bacteria that makes them highly susceptible to contamination and microbial spoilage. Fish are also rich in unsaturated fatty acids that are susceptible to oxidation and the development of off-flavors and rancidity.

The quality of fish and seafood is measured by smell and appearance. Fresh finfish should have a mild, pleasant odor and bright, shiny skin with the scales tightly attached. Fish with the head intact should have clear, bulging eyes and bright red, moist gills. The flesh of fresh fish should be firm and elastic to the touch.

All seafood must originate from a supplier operating under a HACCP plan. Fish must be commercially and legally caught or harvested, except when caught recreationally and approved for sale by the regulatory authority. Ready-to-eat raw, marinated, or partially cooked fish other than molluscan shellfish must be frozen to time and temperature guidelines that meet *FDA Food Code* specifications in order to kill parasites. These guidelines are further described in Figure 4.16. Records must be retained that show how the product was handled.

Shellfish must be purchased from sources approved by the Food and Drug Administration and the health departments of states located along the coastline where the shellfish is harvested. Shellfish transported from one state to another must come from sources listed in the Interstate Certified Shellfish Shippers List. The reason for requiring tight control over molluscan shellfish is to reduce the risk of infectious hepatitis and other foodborne illnesses that may result from eating raw or insufficiently cooked forms of this product. Molluscan shellfish caught recreationally may not be used or sold in retail food establishments.

When received at a retail food establishment, molluscan shellfish should be reasonably free of mud, dead shellfish, and shellfish with broken shells. Damaged shellfish must be discarded.

Molluscan shellfish must be purchased in containers that bear legible source-identification tags or labels (Figure 4.10) fastened to the container by the harvester and each dealer that shucks, ships, or reships the shellstock. Molluscan shellfish tags must contain the following information:

- ▲ The harvester's identification number
- ▲ The date of harvesting
- ▲ An identification of the harvest location or aquaculture site including an abbreviation of the state or country in which the shellfish are harvested
- ▲ The shellfish type and quantity
- ▲ A statement in bold, capitalized type that says, "**THIS TAG IS REQUIRED TO BE ATTACHED UNTIL CONTAINER IS EMPTY AND THEREAFTER KEPT ON FILE FOR 90 DAYS.**"

A shellstock tag must remain attached to the original container until the container is empty. These tags must be held for at least 90 days after the container is emptied. If seafood is suspected of being the source of foodborne illness, the investigating team can use the tags to determine where and when the product was harvested and processed.

Figure 4.10 Example of Shellfish Tag

For display purposes, shellstock may be removed from the tagged or labeled container in which they are received. The shellstock must be placed on drained ice or held in a display container. Care must be taken to protect shellstock from contamination during display.

The identity of the source of shellstock that has been removed from a tagged or labeled container for display must be preserved. This can be accomplished by using an approved record keeping system that keeps shellstock tags or labels in sequence based upon the date when, or dates during which, the shellstock are sold or served. Shellstock from one tagged or labeled container must not be commingled with shellstock from another container before being ordered by the customer.

Fruits and Vegetables

Most fruits and vegetables spoil very rapidly. They continue to ripen even after they are picked. Therefore, they may become too ripe if not properly handled. Microorganisms found in water and soil can also cause fruits and vegetables to spoil. Fruits and vegetables hold their top quality for only a few days.

Some products, like wild mushrooms, may only be used if they have been inspected and approved by a mushroom-identification expert who is approved by the regulatory authority. Beware of fresh mushrooms that are packaged in styrofoam trays and covered with plastic shrink-wrap. Mushrooms have a high rate of respiration that will quickly use up the oxygen inside the package. Unless holes are poked in the plastic wrap that covers the package to permit oxygen inside, oxygen-free conditions may occur that are favorable for the growth of *C. botulinum* bacteria.

Most fresh fruits and vegetables are usually not considered potentially hazardous foods (PHFs). Often the acidity and/or outer skin will prevent the growth of harmful bacteria. However, sprouts and cut melons are now considered PHFs and should be handled in the same manner as other PHFs. Even so, the number of cases of foodborne illnesses linked to these kinds of products has increased in recent years. This is largely due to increased consumption of fresh fruits and vegetables and the emergence of microbes that can cause disease with a low number of organisms. Shiga toxin-producing *Escherichia coli, Shigella spp.*, Hepatitis A, and *Cyclospora spp.* can be infective with only a few cells. Therefore, they do not require a potentially hazardous food to multiply.

Purchase raw fruits and vegetables from approved sources and wash them thoroughly to remove soil and other contaminants before they are cut, combined with other ingredients, cooked, served, or offered for human consumption in a ready-to-eat form. Though not required, whole raw fruits and vegetables may be washed using cleaners. In some cases, an additional anti-microbial rinse (water or bath) is used to reduce the number of microorganisms present on the surface. When these types of chemicals are used, they must meet the requirements in the Code of Federal Regulations (21CFR173.315). The fruits and vegetables should also be rinsed to remove as much of the residues of these chemicals as possible.

Whole raw fruits and vegetables that will be washed by customers before they are eaten do not need to be washed at the establishment before they are sold.

Juice

This group of foods includes the liquid extracted from one or more fruits or vegetables, purees of the edible portions of one or more fruits or vegetables, or any concentrates of such liquid or puree. According to the *FDA Food Code,* this group of foods includes juice as a whole beverage, an ingredient of a beverage, and a puree as an ingredient of a beverage.

Most of the juices sold in retail food establishments are obtained from processors in a prepackaged form. These juices must be obtained from a processor that has a HACCP system in place (See Food Safety Management Programs). In most instances, the processor will pasteurize or otherwise treat the juice to attain 99.999% reduction of the most resistant disease-causing microorganisms of public health significance. The *FDA Food Code* requires juice that is not pasteurized or otherwise treated to bear a warning label that states "WARNING: This product has not been pasteurized and, therefore, may contain harmful bacteria that can cause serious illness in children, the elderly, and PERSONS with weakened immune systems."

Juice packaged in a retail food establishment must:

- ▲ Be treated under a HACCP plan as specified in 8-201.12(B)–(E) of the *2001 FDA Food Code* to attain a 99.999% reduction of the most resistant microorganisms of public health significance, or
- ▲ Bear a warning label that informs customers "This has not been pasteurized and, therefore, may contain harmful bacteria that can cause serious illness in children, the elderly, and PERSONS with weakened immune systems."

Frozen Foods

Frozen products must be solidly frozen when delivered. Check the temperature of frozen foods by placing the sensing portion of a thermometer between two packages. Receiving personnel should also look for signs the product has been thawed and refrozen. Common signs of thawing and refreezing are large ice crystals on the surface of a frozen food and frozen liquids or juices inside the package. Reject frozen foods that are not solidly frozen or show signs of temperature abuse.

Storage of Food

Merely assigning someone to check incoming shipments is not enough. Besides knowing what to look for upon delivery, employees must also know how to store food products safely that are left at the establishment. Foods must be quickly moved to the required storage areas. Employees must use clean carts and dollies when transporting food to prevent contamination. Any products left lying around the establishment by customers who change their minds about a purchase may need to be discarded if perishable. Another problem is a product returned by the customer. It may appear intact, but is questionable. Always remember, "When in Doubt, Throw it Out!"

Stock rotation is a very important part of effective food storage. A First-in, First-out (FIFO) method of stock rotation helps ensure older foods are used first (Figure 4.11). Product containers should be marked with a date or other readily identifiable code to help employees know which product has been in storage longest. When expecting food shipments, always make certain the older stock is moved to the front of the storage area to make room for the newly arriving products.

Figure 4.11 Proper Stock Rotation

Types of Storage

The three most common types of food storage areas in retail food establishments are the:

- ▲ Refrigerator
- ▲ Freezer
- ▲ Dry storage.

Refrigerated storage is used to hold potentially hazardous and perishable foods for relatively short periods of time, usually a few days. Freezer storage is used to hold foods for longer periods of time. Dry storage is typically used to store less perishable items and foods that are not potentially hazardous.

Refrigerated storage slows down microbial growth and preserves the quality of foods. Some common types of refrigerated storage equipment found in retail food establishments are walk-in, reach-in, coffin, upright, under-the-counter refrigerators, and cold display units, among others. A more detailed description of some of these refrigeration systems is contained in Chapter 5. This equipment usually maintains the air temperature in the storage compartment at about 38°F (3° C). Potentially hazardous foods must be stored at 41°F (5°C) or below. Fish and shellfish are especially vulnerable to spoilage and should be stored at colder temperatures ranging from 30°F (-1.1°C) to 34°F (1.1°C). Some fruits and vegetables, such as bananas and potatoes, undergo undesirable chemical changes when they are refrigerated. Therefore, although fruits and vegetables are perishable products, not all types of them should be refrigerated. For perishability, fresh fruits and vegetables requiring refrigeration should be stored at temperatures between 33°F (1°C) and 41°F (5°C). Unpasteurized juices and potentially hazardous fruit and vegetables, like cut melons and sprouts, must be refrigerated at 41°F (5°C) or below.

Refrigerators must be equipped with a temperature-measuring device located to measure the air temperature or a simulated product temperature in the warmest part of the refrigerated unit. Defrost times and cycles should be posted in the appropriate department or on the back of each unit.

Freezer storage is designed to keep foods solidly frozen. Freezer equipment must also be equipped with indicating or recording thermometers to monitor the temperature of the ambient air inside the unit. If your freezer is not frost free, defrost it regularly to ensure proper operation. Wrap frozen foods and transfer them to the refrigerated storage area until the defrosting process is complete.

Some important procedures for cold storage are listed below:

- ▲ Keep refrigerated foods at 41°F (5°C) or below and frozen foods solidly frozen during storage
- ▲ Rotate refrigerated and frozen foods on a First-in, First-out (FIFO) basis and store foods in covered containers that are properly labeled and dated
- ▲ Store foods in refrigerated and freezer storage areas at least 6 inches off the floor and space products to allow the cold air to circulate around them
- ▲ Store raw products under cooked or ready-to-eat foods to prevent cross contamination

- ▲ Keep different species of raw animal foods separate during storage. If limited storage space makes it necessary to store different species in the same area of the refrigerator, store poultry on the bottom shelf, ground beef and pork on the middle shelf, and fish, eggs and other cuts of red meat on the top shelf.

The dry storage area is used to store grocery items that are usually packaged in cans, bottles, jars, and bags (see Figure 4.12). These products must be labeled according to federal regulation and come packaged from approved commercial facilities.

Figure 4.12 Dry Storage Area

The grocery storage area should have a moderate room temperature of 50°F (10°C) to 70°F (21°C) and a relative humidity of 50 to 60% to maximize the shelf life of the foods stored there. The quality of some foods is reduced by sunlight. Therefore, windows are not recommended in dry storage areas. If windows are present, they should be shaded or opaque.

Dry foods should be stored on slatted shelves or skids. Shelves must be at least 6 inches (15 cm.) off the floor, and skids must be easily moveable (Figure 4.13). The 6-inch clearance and easy movement permit thorough cleaning under the shelving and skids. It also discourages insects and rodents from harboring there. Products should be spaced on shelves so air can circulate around them.

Figure 4.13 Store Dry Foods on Slatted Shelving

If it is necessary to transfer bulk food items, such as flour, sugar, and grain foods, into other containers, these containers must be constructed of food-grade materials and equipped with tight-fitting lids. Scoops and other utensils used to remove food from bulk food containers must be constructed of nontoxic, nonabsorbent, and easily cleanable material. These utensils must be equipped with handles and stored in a manner that will allow food employees to grip the handle of the utensil without touching the food with their hands.

Make certain the containers are properly coded, or dated, and labeled with the common name of the food. Containers holding food that can be easily and accurately recognized, such as dry pasta, do not need to be labeled.

Do not use toilet rooms, locker areas, mechanical rooms, and similar spaces for storage of food, single-service items, paper goods, or equipment and utensils. Do not expose foods to overhead water and sewer lines unless the lines are shielded to interfere with potential drips (Figure 4.14).

Figure 4.14 Contamination of Ready-to-Eat Food

Toxic chemicals, such as cleaners, sanitizers, and pesticides, are commonly used and sold in retail food establishments. Most of these products can be poisonous if consumed accidentally. Regardless of the type of chemical, it must be properly labeled. A good label identifies the chemical, provides directions on how to use it safely, and instructs people on what first-aid measures to use in case of accidents.

Improper storage of non-food items can cause chemical contamination. To avoid contaminating food with toxic chemicals, it is important to keep food and non-food items separate during storage (Figure 4.15). Ideally, a separate room should be used to store non-food items. If a separate space is not available, non-food items should be stored below and as far away from food items as possible.

Figure 4.15 Do Not Store Chemicals Near Food

Care should also be taken when storing other miscellaneous items in dry storage areas. Remember, anything that comes in contact with food could lead to food contamination. Items such as packaging materials, food containers, wrapping, meat twine, etc., should all be stored in such a manner to prevent them from being contaminated. Additionally, employees should be discouraged from storing clothing, handbags, and other personal items in grocery storage areas. These items can also be a source of food contamination.

Storage Conditions for Foods

Meat and meat products come in a variety of forms including fresh, frozen, cured and smoked, dried, and canned. Fresh meat can be stored for up to three weeks at temperatures between 28°F (-2°C) and 32°F (0°C) and a relative humidity between 85 and 90%. Cold temperatures extend the shelf life of red meats by slowing down the growth of bacteria that cause spoilage. The high relative humidity prevents excessive drying and shrinkage.

Frozen meats can be stored for several months when held at 0°F (-18°C) or below. Frozen meats must be wrapped in moisture-proof paper to prevent them from drying out. Packaging for frozen foods should also be strong, flexible, and protect against light.

Poultry can be safely stored at temperatures between 28°F (-2°C) and 32°F (0°C) for short periods of time. A relative humidity of 75 to 85% is recommended, as excessive humidity causes sliminess due to excessive bacterial growth. Poultry should be wrapped carefully to prevent dehydration, contamination, and loss of quality. Frozen poultry and poultry products can be stored for four to six months when held at 0°F (-18°C) or below.

Whole shell eggs will keep fresh for up to two weeks when stored at 41°F (5°C) or below. It is, however, recommended to store eggs at 34°F (1°C) to 38°F (3°C) to maintain optimum quality. Egg quality deteriorates rapidly at room temperature. Eggshells are porous and odors can be absorbed very easily. Keep eggs covered and store them away from onions and other foods that have a strong odor. Discard eggs that are dirty or cracked. Always make sure to wash your hands after handling whole shell eggs.

Egg products such as whole eggs, egg whites, and yolks are pasteurized to destroy *Salmonella* bacteria. Fresh egg products should be stored at 41°F (5°C) or below. Store frozen eggs at 0°F (-18°C) or below and keep them frozen until time for defrosting. Refrigerate dried egg products or keep them in a cool, dry place. Once dried eggs have been reconstituted, they are considered potentially hazardous and must be stored out of the temperature danger zone.

Milk is one of the most perishable foods handled in a food establishment. However, properly pasteurized milk that has not been recontaminated and is held at 41°F (5°C) or less will keep up to ten days or longer. The optimal storage temperature for fluid milk is 34°F (1°C) to 38°F (3°C), and the shelf life of milk is shortened significantly at higher storage temperatures. For example, milk held at 60°F (16°C) would stay fresh less than one day. Milk also picks up odors from other foods. Therefore, store milk in an area away from onions and other foods that give off odors.

Fish and shellfish are more perishable than red meats, even when refrigerated or frozen. There are many bacteria in the surface slime and digestive tracts of living fish. When a fish is killed, bacteria attack the tissue of the fish. This bacteria lives on cold-blooded fish at low water temperatures. Therefore, they adapt well to the cold and continue to grow even under refrigerated conditions. The shelf life of fresh fish depends on many factors including: species, season of the year, physical condition (whether filleted or dressed), and manner of handling.

Most shellfish are even more perishable than finfish. The shells of live clams and oysters should be closed or close when tapped. Lobsters and crabs should be kept alive until they are cooked or frozen. Otherwise, they lose quality in a day or less. An easy way to determine if a lobster is still alive is to touch the lobster's eye and watch for movement.

Products stored in ice should be placed under cooked or ready-to-eat foods to prevent cross contamination. Fish and shellfish must be kept at 41°F (5°C) or below for food safety reasons. Storage conditions to maintain optimum quality range from 30°F (-1°C) to 34°F (1°C).

The ideal storage conditions for fresh fruits are temperatures between 33°F (1°C) and 41°F (5°C), a relative humidity of about 80%, and shaded from light. During respiration and ripening, fresh fruits give off carbon dioxide and water. Because of this, they may become wilted and lose flavor. Proper air circulation in the refrigerated storage area is necessary to maintain freshness and firmness. Inadequate ventilation leads to spoilage and deterioration. Regularly inspect stored fruits. Discard any that begin to spoil.

Vegetable storage requires low temperatures and high humidity to preserve texture, tenderness, flavor, color, and nutritive content. To retain top quality, vegetables should be stored between 33°F (1°C) and 41°F (5°C) with a relative humidity of 85 to 95%.

Food Preparation, Handling, and Service

Preparation, handling and service of food in retail food establishments involve one or more steps. In convenience stores, ready-to-eat products, such as sandwiches and salads, are sold to customers with minimal preparation. In larger retail food establishments, foods are prepared in consumer-sized portions using commercially available equipment for cutting, grinding, slicing, cooking, and cooling. These foods are commonly referred to as "meal solutions" or "home meal replacements."

In recent years, many retail food establishments have expanded into food processing-type operations, and some establishments are altering the atmospheres within food packages or applying federally regulated food additives as methods of food preservation. Some examples of food processing operations now being conducted in retail food establishments include cook-chill, vacuum packaging, smoking and curing, and custom processing of animals. The *FDA Food Code* recommends a HACCP plan (see Chapter 10) be developed and approved prior to conducting these types of food processing operations within a retail food establishment. Additional information about some of these processes is presented later in this chapter.

Regardless of how many steps are involved in food production and service, foodborne illness prevention requires effective food safety measures that ensure good personal hygiene and avoid cross contamination and temperature abuse. Food safety strategies aimed at addressing these practices will be discussed in the remainder of this chapter.

During preparation, an important technique that can be used to promote food safety is "small batch" preparation. Food preparation is usually done at room temperature. This is several degrees into the temperature danger zone. Therefore, you must limit the amount of time the food is in the danger zone by working with small and manageable amounts of potentially hazardous ingredients. In addition, some preparation procedures may require ingredients to be pre-cooled prior to preparation (i.e., ingredients used in the preparation of chicken salad).

Ingredient Substitution

For meal solutions and home meal replacements, there is normally a recipe for the products prepared in the retail food establishment. The recipe usually includes a list of ingredients and instructions for how to prepare, store, and label the food item. When one or more of the original ingredients is not available for a recipe, other ingredients may be substituted so the food item can still be prepared and sold. However, all ingredient substitutions should be identified in the recipe beforehand. Ingredient substitutions must never compromise the safety of the food and should not be allowed unless they are identified and allowed in the recipe.

Avoiding Temperature Abuse

Temperature and time are the most common critical control points employed during preparation and service. Monitoring and controlling food temperatures are extremely effective ways to minimize the risks of foodborne illnesses. Thermometers are used for stored, cooked, hot-held, cold-held, and reheated foods. Various types of thermometers are used throughout the establishment. Some of these instruments were described and shown in Figure 3.4 (see thermometer on page 81).

Before using a thermometer, make sure it is clean, sanitary, and properly calibrated. Always insert the "sensor" portion or probe stem of the thermometer into the thickest part of the food. In most instances, this will be at the center of the food product or container.

Freezing

Most disease-causing microorganisms do not grow at temperatures below 41°F (5°C). However, some spoilage microorganisms can grow well below 41°F (5°C). Freezing foods provides a good alternative to maintaining quality and safety for longer storage periods not possible with refrigeration or room temperature storage. Frozen foods must be kept solidly frozen for optimum quality.

Although bacteria are generally not destroyed by freezing, parasites can be killed if foods are frozen at the proper temperature for the proper length of time. Figure 4.16 presents the parasitic destruction guidelines that have been established for destroying parasites in raw-marinated and marinated, partially cooked fish. Yellowfin, bigeye, bluefin-northern, bluefin-southern, and certain other species of tuna may be served or sold in a raw, raw-marinated, or partially cooked ready-to-eat form without freezing.

Food should be frozen throughout to -4° F (-20°C) and held for 7 days in a freezer,

or

Food should be frozen throughout to -31°F (-35°C) using a blast chiller, and held at that temperature for 15 hours.

Figure 4.16 Guidelines for Parasitic Destruction
(Source: *FDA Food Code*)

When measuring the temperature of packaged frozen foods, place the sensing portion of a thermometer between two fully frozen packaged foods. Monitor the temperature of freezer units regularly to make sure they are working properly. Always remember measuring the air temperature in the freezer is never as effective as measuring actual food product temperature.

Thawing

Freezing prevents microbial growth in foods, but usually does not destroy all microbes. Improper thawing provides an opportunity for surviving bacteria to grow to harmful numbers and/or produce toxins. The most common and acceptable methods for thawing foods include: in a refrigerator, in a microwave oven followed by immediate cooking, under cool running water [less than 70°F (21° C)], and as part of the cooking process (Figures 4.17 and 4.18).

Figure 4.17 Safe Ways to Thaw Food

The preferred method of thawing is in the refrigerator, since foods thawed in this manner would not have the opportunity to be in the temperature danger zone. This method requires good planning and adequate refrigeration space. It can take two or three days and sometimes longer to thaw large food masses such as turkeys, roasts, and hams. When using this method to thaw foods, you must plan ahead.

Foods thawed in the microwave must be cooked immediately after thawing. Cooking can be done in the microwave oven or with conventional cooking equipment. Do not thaw foods in a microwave oven and then store them in a refrigerator before they are cooked. The thawing process may initiate microbial activity that could reach dangerous levels during refrigerated storage.

Potentially hazardous foods, like turkey and frozen shrimp or scallops, can be thawed when they are submerged under cool [less than 70°F (21°C)] running water. It is desirable to thaw foods as quickly as possible when using this method. The *FDA Food Code* requires raw animal foods to be thawed in less than four hours including the time it takes for preparation for cooking or to lower the food temperature to 41°F (5°C) under refrigeration. Thawed portions of ready-to-eat foods should not be allowed to rise above 41°F (5°C) when using the cool water thawing process.

- ▲ Under refrigeration that maintains the food temperature at 41°F (5°C) or below
- ▲ Completely submerged under running water:
    - ▼ at a water temperature of 70°F (21°C) or below
    - ▼ with enough water velocity to remove contaminants from the surface of the food
    - ▼ for a period of time that does not allow thawed portions of ready-to-eat foods to rise above 41°F (5°C)
    - ▼ for a period of time that does not allow thawed portions of a raw animal food requiring cooking to be in the temperature danger zone for more than a total time of 4 hours
- ▲ As a part of the cooking process
- ▲ Use any procedure (i.e., microwave oven) that thaws a portion of frozen ready-to-eat food that is prepared for immediate service in response to an individual customer's order.

Figure 4.18 Guidelines for Thawing Food
(Source: *FDA Food Code*)

Retail food establishments will sometimes use a slacking (defrosting) process to moderate the temperature of foods prior to cooking or re-heating. During the slacking process, foods can be defrosted under refrigeration that maintains the food at 41°F (5°C) or less or at any temperature if the food remains frozen. The slacking process is typically used with previously block-frozen food such as spinach.

Under no circumstances should foods be thawed or slacked at room temperature. Room temperature thawing puts foods in the temperature danger zone–the very thing you don't want to have happen. When foods are thawed at room temperature, the outer surface of the food thaws first and will soon reach room temperature. Microbial growth occurs very quickly at room temperature.

Cold-Holding

Most harmful microorganisms start to grow at temperatures above 41°F (5°C). Therefore, it only makes sense to hold potentially hazardous and perishable foods in cold storage at 41°F (5°C) or below.

Some bacteria, such as *Listeria monocytogenes,* can grow at temperatures below 41°F (5°C). You should monitor the temperature of cold-held foods routinely throughout the day to ensure safe temperature requirements are being met.

Refrigerators, display cases, and cold service bars differ in their capacity to keep foods cold. In all instances, the air inside the refrigerated compartment will need to be a few degrees colder than 41°F (5°C) to ensure cold foods are held at 41°F (5°C) or below.

When displaying raw potentially hazardous foods like meats [Figure 4.19(a)], be sure they are maintained at less than 41°F (5°C). Usually the temperature of the refrigerator or display case should be at least 3°F (1.5°C) to 5°F (3°C) below the desired cold-holding temperature. Prior to packaging and display, meat-cutting operations should be done in a refrigerated processing room with an ambient air temperature between 45°F (7°C) and 55°F (13°C). This temperature allows the employees to be more comfortable and slows the growth of any bacteria that may be present on the meat.

Fish and seafood products should be placed on ice [Figure 4.19(b)] to prevent the growth of disease-causing and spoilage bacteria. The ice used for this purpose must be made with potable water. It must be produced by an ice machine that is maintained in a clean and sanitary condition. The ice must be transported to display cases in containers that are made of materials approved for food-contact surfaces. (See Chapter 5 for details.) When the ice melts, the liquid must be drained away from the food to prevent contamination. Ice that has been in contact with fish and shellfish must be considered contaminated and cannot be used for any other purpose.

Cold, ready-to-eat, potentially hazardous foods include deli meats, potato and macaroni salads, chicken and seafood salads, cooked shrimp, and similar items. The *FDA Food Code* recommends when these products are prepared and held refrigerated for more than 24 hours, they be clearly marked at the time of preparation to indicate "sell-by" or "use-by" for the customer. The acceptable storage time for these types of products is seven calendar days when held at 41°F (5°C) or below.

a. Fresh meat

b. Fresh fish

Figure 4.19 Cold Display of Raw, Potentially Hazardous Foods

Ready-to-eat salads (i.e., chicken, seafood, potato, or pasta) must be prepared in a way that will protect them from temperature abuse. This begins with pre-chilling all ingredients in a refrigerator before mixing and storing. By pre-chilling the ingredients, you are able to keep the temperature of the finished product as cold as possible. For example, when making potato salad you should make sure the cooked potatoes are properly cooked and cooled before prepping the salad. Other ingredients such as celery, mayonnaise, and mustard should also be stored overnight in the refrigerator to make sure they are cold prior to mixing. It is also a good idea to prepare foods in small batches in order to assure the food is not in the temperature danger zone too long. Large tubs of potato salad may take one to two days to cool in the refrigerator.

Frozen, ready-to-eat, potentially hazardous foods should be sold within 24 hours after thawing whenever possible. When large amounts of food are removed from the freezer, it should be marked to indicate the date by which the food must be sold. Food must be used within seven calendar days or less after the food is removed from the freezer, minus the time before freezing that food is held refrigerated, if the food is maintained at 41°F (5°C) or less before freezing.

When displaying ready-to-eat potentially hazardous foods like prepared salads and luncheon meats [Figure 4.20(a)] be sure the refrigerator unit can maintain a safe cold-holding temperature. This may be more difficult in open top and open front refrigerated display cases that do not have doors. Refrigerated display cases have a "safe load line" [Figure 4.20(b)]. This line indicates the level below which foods must be stored to ensure the food is held at the proper temperature. It is also important to store foods so that the discharge or return air vents are not blocked.

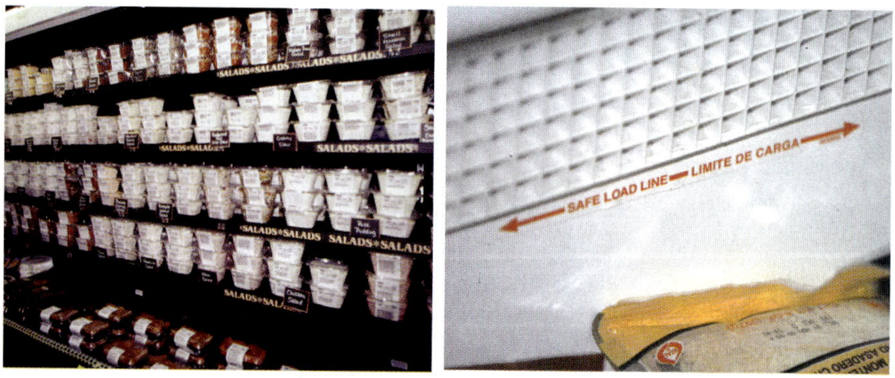

a. Prepared salads b. The safe load line

Figure 4.20 Cold Storage of Ready-to-Eat Foods

A refrigerated, ready-to-eat, potentially hazardous food prepared and held in a retail food establishment for more than 24 hours, or originating from an original container prepared and packaged by a food processing plant and opened at the retail food establishment must be discarded if it:

- ▲ Exceeds the prescribed time and temperature requirements
- ▲ Is in a container or package that does not bear a date or day or
- ▲ Is marked with a date or day that exceeds the time and temperature combinations described above.

Cooking

Raw foods, especially those of animal origin, contain harmful microorganisms. Foods such as meat, poultry, fish, seafood, eggs, and unpasteurized milk should not be prepared and served raw or rare. Retail food establishments that choose to serve raw foods increase their risk of having a foodborne illness. Practices such as using raw eggs to make a Caesar salad or raw oysters on the half shell increase the risk of a foodborne illness. Raw animal foods need to be cooked to the proper temperatures to be safe. If foods are cooked within a foodservice operation of a retail food establishment, it is important employees know how to do it properly. Many retail establishments sell raw PHFs with the intent the customer will properly cook it. For these situations, safe cooking information and guidance should be provided for the customer. This can be done with the use of labels (i.e., Safe Food-Handling Instructions), cooking recipe cards, or store placards. Many retail food establishments now offer food safety brochures to help the consumer.

The purpose of cooking is to make food more palatable by changing its appearance, texture, and aroma. Cooking also heats the food and destroys harmful microorganisms that may be found in and on the product. The destruction of disease-causing microorganisms is a phenomenon involving a direct relationship between time and temperature. Requirements for cooking a particular food item should include a final internal temperature as well as a prescribed length of time at that temperature. For example, a rare roast beef could be safely cooked to 130°F (54°C) for 112 minutes or 140°F (60°C) for 12 minutes.

Most foods are cooked using conventional ovens, microwave ovens, fryolators, steamers, and rotisseries. Because heat transfer can be different depending on the heating source, final temperature requirements have been set for conventional oven cooking and microwave cooking. Cooking guidelines for various kinds of potentially hazardous foods are presented in Figure 4.21.

Rare beef roasts have the lowest internal temperature requirements. This is because pathogens are typically found on the surface of the roast. When an internal temperature of 130°F (54°C) for 112 minutes or 140°F (60°C) for 12 minutes is reached, the surface temperature of the food is much higher. This heat will destroy any pathogens that may have been present on the surface of the roast. Animal foods such as eggs, fish, and beef (other than roasts) should be cooked to an internal temperature of at least 145°F (63°C) and held at that temperature for at least 15 seconds before serving.

Cook ground beef, ground pork, and ground game animal products to 155°F (68°C) and hold for at least 15 seconds before serving. Pork and game animals need a higher cooking temperature due to the possible presence of the trichinosis worm and other parasites.

Poultry and stuffed meats should be cooked to an internal temperature of 165°F (74°C) and held for at least 15 seconds before serving.

Casseroles and other foods that contain a combination of raw ingredients such as meat and poultry must be cooked to a final temperature that coincides with the highest risk food. In this case, 165°F (74°C) is required to destroy pathogens that may be found in the poultry. Food mixtures, such as chili and beef stew, must be cooked to 165°F (74°C) to assure proper destruction of disease-causing agents.

Food Type	Minimum Internal Temperature	Minimum Time Held at Internal Temperature Before Serving
Beef Roast (rare)	130°F (54°C) 140°F (60°C)	112 min. 12 min.
Eggs, Beef and Pork (other than roasts), Fish	145°F (63°C)	15 sec.
Ground Beef, Ground Pork, and Ground Game Animals	155°F (68°C)	15 sec.
Beef Roast (medium), Pork Roast, and Ham	145°F (63°C)	4 min.
All Poultry, Stuffed meats	165°F (74°C)	15 sec.

Note: When microwave cooking, heat raw animal foods to a temperature of 165°F (74°C) in all parts of the food.

Figure 4.21 Cooking Guidelines for Potentially Hazardous Foods
(Source: *FDA Food Code*)

While not usually potentially hazardous foods, fruits and vegetables that are cooked should reach an internal temperature of 140°F (60°C) or above.

Potentially hazardous foods prepared from ingredients, such as reconstituted foods and canned tuna, which have been held at room temperature, must be cooled to 41°F (5°C) or less within four hours.

When cooking foods in the microwave oven, the distribution of heat is often uneven. Stirring and rotating the food during the cooking process will enable heat to be distributed more evenly. The *FDA Food Code* requires raw animal foods cooked in a microwave oven to be heated to 165°F (74°C) in all parts of the food. As a common practice, foods cooked in a microwave oven should be allowed to stand covered for two minutes before serving to allow the heat inside the product to disperse more evenly.

The internal temperature of foods must be measured during the cooking process. For most foods, the internal temperature will be measured by inserting the probe of a thermometer or thermocouple into the center or thickest part of the food mass. This will give you an accurate reading of the internal temperature of the product.

Proper cooking is extremely important for the preparation of safe food. If a potentially hazardous food is cooked to a safe temperature, bacteria and other harmful organisms will be destroyed. You should remember from Chapter 2 normal cooking temperatures may not destroy bacterial spores and toxins. Therefore, proper handling of foods that may contain these agents is required at every stage of the food flow.

Cooling

Improper cooling is one of the leading contributors to foodborne illness in retail food establishments. Foods are in the temperature danger zone during cooling and there is no way to avoid it. After proper cooking, potentially hazardous foods need to be cooled from 140°F (60°C) to 41°F (5°C) as rapidly as possible. The *FDA Food Code* recommends hot foods, not used for immediate service or hot display, be cooled from 140°F (60°C) to 70°F (21°C) within two hours, and from 140°F (60°C) to 41°F (5°C) or less within six hours.

Potentially hazardous foods prepared from ingredients, such as reconstituted foods and canned tuna, which have been held at room temperature, must be cooled to 41°F (5°C) or less within four hours.

Large quantities of food and foods that have a thick consistency take a long time to cool. For instance, it can take 72 hours or more for the center of a 5-gallon stockpot of steamed rice to cool down to 41°F (5°C) when taken hot from the stove and placed in a refrigerator. Reach-in refrigerators should not be used to cool foods. It is very difficult to cool foods quickly and the temperature in the reach-in refrigerator can rise considerably.

Foods must pass through the temperature danger zone as quickly as possible. Figure 4.22 lists some of the more commonly used methods for reducing cooling time.

- ▲ Blast chillers
- ▲ Walk-in coolers, loosely covered
- ▲ Use containers that facilitate heat transfer (stainless steel)
- ▲ Transfer food into shallow pans that will allow for a product depth of 3" or less
- ▲ Transfer food into smaller containers
- ▲ Place container of hot food in an ice water bath
- ▲ Stir food while cooling
- ▲ Use cooling paddles to stir the food
- ▲ Add ice as an ingredient directly to a condensed food.

Figure 4.22 Methods to Reduce Cooling Time for Food

Photographs of the two most commonly used methods of cooling are presented in Figure 4.23. As with cooking, you must monitor times and temperatures for foods that are cooling. When large masses of food are broken down into smaller portions, cooling is more rapid.

For example, a large turkey can be sliced into smaller pieces, or creamed corn can be transferred into shallow pans to facilitate cooling. Foods cool faster in food-grade metal containers than in food-grade plastic containers.

When using shallow pans to cool liquid food, the pan should be only 4 inches high. The depth of foods cooled in a shallow pan should be less than 3 inches for liquids (soup) and less than 2 inches for viscous products (chili or beef stew) [Figure 4.23(b)].

a. Ice bath

b. Place food in shallow pans 2 to 3" deep

Figure 4.23 Safe Cooling Methods

Cooling time is greatly reduced when foods are stirred. Foods stirred every 10-15 minutes cool far more rapidly than unstirred foods. An ice water bath also helps to remove heat from foods.

An ice water bath involves putting the container of hot food inside a larger container and surrounding it with ice to promote cooling. It is sometimes desirable to combine cooling methods, such as stirring a food after putting the container into an ice water bath.

You may also fill a large pot with ice and water. Then, insert the container of hot food into the ice water and stir the product. Keep the water level below the rim of the food container. In some instances, ice from a potable water supply can be added as a "final ingredient" in preparing hot items from condensed foods. Large quantities of soup or broth can be prepared in this way.

Always use a thermometer to verify foods are cooling properly. Never assume any one method is working without checking the temperature and time foods take to cool. Always depend on the thermometer reading with any of the methods you use to cool food.

An example of the different methods that may be used to cool a large mass of chili are presented in Figure 4.24. The chili was cooled in a large stock pot, in shallow pans (with a product depth of 2 inches), and using an ice water bath. Remember, it is the time between 140°F (60°C) to 41°F (5°C) that is important. Notice how long it takes for the chili to cool in the large stockpot. Even after nine hours, the chili was not properly cooled. Use of shallow pans or an ice water bath were both effective methods for cooling. Using shallow pans, the chili could be cooled to the required final temperature in less than three hours. When the ice water bath was used, the chili was cooled to below 41°F (5°C) in less than two hours. Always measure the temperature of foods during cooling to ensure the cooling method is working well.

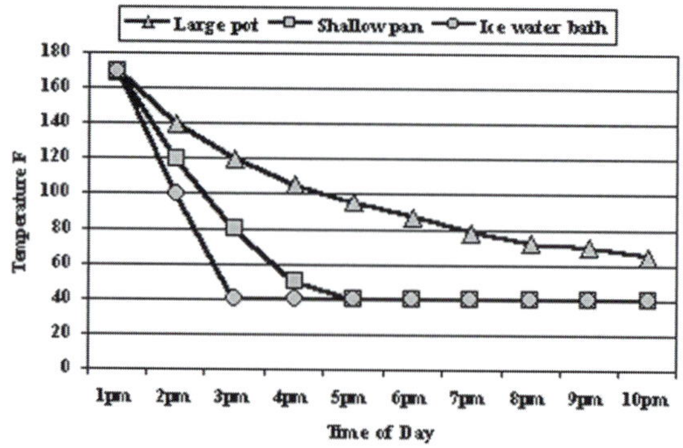

Figure 4.24 Different Cooling Methods for Chili

Hot-Holding and Reheating

All potentially hazardous foods that have been cooked and are intended to be held hot (not cooled, stored, and reheated) must be maintained at 140°F (60°C) or above (Figure 4.25). Hot-holding is also required when hot potentially hazardous foods are delivered to sites away from the retail food establishment. During hot-holding, never add fresh product to existing product and always work in small batches. This is a good approach to reduce food contamination, assure good temperature control, and maintain product appeal.

Figure 4.25 Hot-Holding—Store Hot-Held Foods at 140°F (60°C) or above

All potentially hazardous foods meant to be held hot, which have been cooked and then cooled, must be reheated to at least 165°F (74°C) within two hours. Foods can be heated more quickly when heated in smaller quantities or by using pre-heated ingredients. Stir food frequently during reheating to reduce the time needed to reheat. For quality and safety reasons, foods should be reheated as few times as possible. Temperature and time are often used together to eliminate microbes or to prevent them from growing in foods.

While temperature is usually the most important factor in controlling microbes, there are some situations where controlling time can also be used. That is why there is an allowance of four hours in the temperature danger zone. The use of time as a control factor is most applicable for prepared ready-to-eat foods that are being stored or displayed just prior to sale. Sliced pizza and fried chicken are good examples. If these potentially hazardous foods were hot-held at 140°F (60°C) or above, the food may dry out and the quality may deteriorate very quickly. Remember, cooking temperature is critical in making the food safe to eat.

Ready-to-eat foods that are commercially prepared and packaged and from a food processing plant under regulatory inspection should be free of harmful microorganisms. Therefore, these foods must only be reheated to a temperature of at least 140°F (60°C) for hot-holding [rather than a minimum of 165°F (74°C)].

To maintain quality and safety, some retail food establishments may hold potentially hazardous foods in the temperature danger zone for a short period of time. Most establishments will limit this amount of time to 20-30 minutes. In these instances, it is critical to achieve the required cooking temperature and the amount of time foods are held in the temperature danger zone must be carefully monitored and recorded. The procedures and monitoring required when using time as a method of control varies. Consult with your local regulatory authority for requirements in your jurisdiction.

Reduced Oxygen Packaging (ROP)

During the past decade, some retail food establishments have expanded into **reduced oxygen packaging** (ROP) to increase the shelf life of foods. The term ROP is defined as any packaging method that reduces the amount of oxygen in a sealed package. Oxygen may be removed or it may be replaced with another gas. ROP is an inclusive term and can include other processing and packaging options such as:

- ▲ Cook-chill – a process that uses a plastic bag filled with hot cooked food from which air has been forced out and which is closed with a plastic or metal crimp.
- ▲ Controlled Atmosphere Packaging (CAP) – a system which maintains the desired atmosphere within a package throughout the shelf life of a product by the use of agents to bind or scavenge oxygen or a small packet containing compounds to emit a gas.
- ▲ Modified Atmosphere Packaging (MAP) – a process that employs a gas flushing and sealing process or reduction of oxygen through respiration of vegetables or microbial action.
- ▲ Sous Vide – a process where fresh raw foods are sealed in a plastic pouch and the air is removed by vacuum. The pouch is cooked at a low temperature and rapidly cooled to 38°F (3°C) or below or frozen.
- ▲ Vacuum Packaging – reduces the amount of air from a package and hermetically seals the package so a near-perfect vacuum remains inside.

Some of the most widely recognized benefits of Reduced Oxygen Packaging are:

- ▲ It creates a largely oxygen-free environment that prevents the growth of aerobic (see Chapter 2) bacteria, yeast, and molds that are largely responsible for the off odors, slime, texture changes, and other forms of spoilage
- ▲ It prevents chemical reactions that can produce off odors and color changes in foods
- ▲ It reduces product shrinkage by preventing water loss.

In spite of its benefits, the use of ROP with some foods can increase safety concerns. Unless potentially hazardous foods are protected inherently, simply placing them in ROP without regard to microbial growth will increase the risk of foodborne illnesses. Some products in ROP contain no preservatives and frequently do not possess any intrinsic inhibitory barriers (such as pH, A_w, or salt concentrations) that either alone or in combination will inhibit microbial growth. Thus, product safety is not provided by natural or formulated characteristics.

The anaerobic environment created by ROP provides the potential for growth of *Clostridium botulinum* and other important pathogens. Some of these organisms, including *Listeria monocytogenes,* can grow slowly at temperatures below 41°F (5°C). Another safety concern of ROP foods is because spoilage bacteria are destroyed there may no longer be telltale signs signaling the product is no longer fit to eat.

The *FDA Food Code* provides very specific requirements for retail food establishments that desire to use ROP technology. A summary of the requirements are presented below:

- ▲ The retail food establishment must have a HACCP plan in place for the ROP operation that details how the seven principles (see Chapter 10) are incorporated into the operation's food safety management system.
- ▲ Because *Clostridium botulinum* is a significant health hazard, the establishment is required to have two microbial growth barriers. Time/temperature are commonly used as one barrier, but they need to be coupled with the use of pH, A_w, or product formulation to assure *Clostridium botulinum* will not grow inside the package.
- ▲ All potentially hazardous foods in ROP which rely on refrigeration as a barrier to microbial growth must be maintained at 41°F (5°C) or below.
- ▲ The refrigerated shelf life is to be no more than 14 days from packaging to consumption or the original manufacturer's "sell by" or "use by" date, whichever comes first.
- ▲ Packages must be prominently and conspicuously labeled on the principal display panel with the instructions to maintain the food at 41°F (5°C) or below and to discard refrigerated food if within 14 days of its packaging it is not served for on-premises consumption or consumed if served or sold for off-premises consumption.
- ▲ Each container must bear a "use-by" or "sell-by" date. This date cannot exceed 14 days from retail packaging or repackaging without a variance granted by the regulatory authority.
- ▲ Employees responsible for the ROP process must receive training that will enable them to understand the key components of the process, the equipment that is used, and the specific procedures that must be followed to ensure critical limits identified in the HACCP plan have been met. The training program must also outline the employees' responsibilities for monitoring and documenting the process and describe what corrective actions they must take when critical limits are not met.

Some products cannot be packed by ROP unless the retail food establishment is approved for the activity and inspected by the appropriate regulatory authority. These products include raw and smoked fish, soft cheeses (ricotta, cottage cheese, and cheese spreads), and combinations of cheese and other ingredients such as vegetables, meat, or fish. Contact your local regulatory authority to obtain a complete set of guidelines for ROP foods and to seek a variance for all food manufacturing/processing operations that are based on the prior approval of a HACCP plan.

Food Handling

Employees must practice good personal hygiene when handling food. This starts with a clean uniform and an effective hair restraint. Food handlers in all departments should avoid touching food with their bare hands. They can use tongs, serving spoons, disposable gloves, or deli tissue when handling meats, cheeses, prepared salads, or when making sandwiches. Employees must hold serving utensils by the handle only, and they must never touch the part of the utensil that comes into contact with food (Figure 4.26). A single utensil should be used for each food item, and the utensil should be stored in the food between uses. Always store serving utensils in a way that permits the employee to grab the handle without touching the food.

Figure 4.26 Handle Eating Utensils and Ice Properly

As a rule, any food that has been served or sold to and in the possession of a customer may not be returned and offered for service or sale to another customer. Two acceptable exceptions to this rule are:

▲ A container of a nonpotentially hazardous food (i.e., a narrow-neck bottle of catsup or steak sauce) that is dispensed in a way that protects the food from contamination and the container is closed between uses

- A nonpotentially hazardous food such as crackers, salt, or pepper that is in an unopened, original package and is maintained in a sound condition.

Finally, employees must remember to wash their hands after touching soiled equipment, utensils, and cloths. Never dry hands on aprons and multi-use towels. If disposable gloves are used, a fresh pair should be put on immediately prior to handling any food products.

Discarding or Reconditioning Food

A food that is unsafe, adulterated or not honestly presented shall be reworked or reconditioned using a procedure that has been approved by the regulatory authority in the jurisdiction, or it must be discarded.

Food must be discarded if it is not from an approved source or has been contaminated by food employees, consumers, or other persons via soiled hands, bodily discharges (i.e., coughs and sneezes), or other means.

Ready-to-eat foods must be discarded if they have been contaminated by an employee who has been restricted or excluded as described in Appendix C of this book.

Refilling Returnable Containers

The practice of refilling returnable containers is not a process commonly associated with retail food establishment. However, if a take-home food container is returned to a retail food establishment, it may not be refilled with a potentially hazardous food at the establishment. Take-home food containers may be refilled with a food that is not potentially hazardous if the container is properly cleaned before being refilled. Except for food-specific beverage containers, returned empty containers intended for cleaning and refilling with food must be cleaned and refilled in a regulated food processing plant.

A food-specific container for beverages may be refilled at a food establishment if:

- Only a beverage that is not a potentially hazardous food is dispensed into the container
- The design of the container and of the rinsing equipment and the nature of the beverage, when considered together, allow effective cleaning of the container at home or in the retail food establishment

- Facilities for rinsing before refilling returned containers with fresh, hot water that is under pressure and not recirculated are provided as part of the dispensing system
- The consumer-owned container returned to the retail food establishment for refilling is refilled for sale or service only to the same consumer
- The container is refilled by an employee of the retail food establishment or the owner of the container if the beverage system includes a contamination-free transfer process that cannot be bypassed by the container owner.

Personal take-out beverage containers, such as thermally insulated bottles, nonspill coffee cups, and promotional beverage glasses, may be refilled by employees or the consumer if the refilling process will protect the food and food-contact surface of the container from contamination.

Self-Service Bar

Self-service salad and hot food buffet bars are very popular in retail food establishments. They offer convenience and a wide range of selections for customers (Figure 4.27). The two most important food safety goals for this type of operation are to:

- Protect foods from contamination by customers
- Keep foods out of the temperature danger zone.

Figure 4.27 Self-Service Bar

A properly installed sneeze guard protects the food from contamination by your customers and is required in most jurisdictions (Figure 4.28).

Raw animal foods, such as oysters, must never be placed on self-service bars. Keep hot-held potentially hazardous foods at 140°F (60°C) or above and cold foods at 41°F (5°C) or below. Employees must monitor food-holding temperatures frequently in these operations.

Figure 4.28 Don't Forget the Sneeze Guard

Use clean and sanitized utensils in a self-service bar and replace any utensils that become contaminated or soiled. Employees and customers must hold utensils by the handle to avoid bare hand contact with foods in containers.

Use only one utensil for each food item, and store it in the food between uses. Resting utensils on saucers beside containers is not a desirable practice.

Self-service bars tend to be very busy at peak selling times. It is important to assign responsibility to this area. Assign a properly trained food handler to monitor and maintain the self-service salad or buffet bar. The employee should clean the self-service bar throughout the day to prevent contamination and cross contamination. If customers are allowed to visit the self-service bar more than once, they must be given a clean plate or bowl for each trip. This will reduce the risk of contaminating food on display at the bar. Beverage cups and glasses may be reused to get refills.

Checkout and Bagging

The last step in the flow of food in retail food establishments takes place at the checkout counter (Figure 4.29). Checkers and baggers must work carefully to prevent contamination and cross contamination. They must practice good personal hygiene and not allow their hands to transfer bacteria from raw animal foods to cooked and ready-to-eat foods. Juices from raw meat, poultry, fish and shellfish can contaminate produce and other ready-to-eat foods. These juices must be routinely wiped off conveyor belts and employee hands through proper hand washing.

Figure 4.29 Keep Checkout Counters Clean and Sanitary

Baggers must be careful not to put raw meat, poultry and fish in the same bags with fresh fruits and vegetables. In addition, non-food items, such as detergents, pesticides, and light bulbs, must not be put in the same bag with foods to avoid contamination with chemical and physical hazards. Freezer bags should also be used to separate frozen goods.

Sometimes customers change their mind about purchasing a particular item. Whenever these are potentially hazardous foods, the product must be put back into the proper storage or display environment as quickly as possible.

Temporary Facilities and Mobile Food Facilities

A temporary food establishment (TFE) is defined by the *FDA Food Code* as a food establishment that operates for a period of no more than 14 consecutive days in conjunction with a single event or celebration. TFEs may operate either indoors or outdoors and often have limited physical and sanitary facilities available.

Some food establishments are offering food by way of temporary and mobile facilities.

A variety of foods can be prepared and served from temporary stands and trailers. Some examples include sandwiches, pizza, BBQ ribs, gyros, corn on the cob, confections (i.e., cotton candy, funnel cakes, or elephant ears), and beverages.

Mobile facilities are trucks and trailers used to cater events located away from the establishment. The extent of food items offered through catering operations is practically unlimited.

a. Mobile b. Temporary

Figure 4.30 Examples of Mobile and Temporary Food Facilities

The same food safety practices that are employed in other areas of the establishment must also be applied at temporary, mobile facilities, and in-store product demonstrations. In particular, food must be protected from:

- ▲ Temperature abuse
- ▲ Infected employees who practice poor personal hygiene and use improper food-handling practices
- ▲ Contamination and cross contamination.

To avoid temperature abuse, keep hot foods hot [140°F (60°C)] or above and cold foods cold [41°F (5°C)] or below. Cooking equipment must be capable of heating potentially hazardous foods to the required internal temperature. For example, hamburgers and other ground meats must be cooked to an internal temperature of 155°F (68°C) and poultry must be cooked to 165°F (74°C). When reheating foods, be sure the product temperature reaches 165°F (74°C) within two hours. Do not use slow cooking devices such as crock pots, steam tables, and sterno-like fuel canisters to cook foods. These devices are designed to keep hot foods hot, but are not intended to cook foods to temperatures that will kill disease-causing organisms. Foods that require refrigeration must be held at 41°F (5°C) or below until sold or served. A thermometer or equivalent temperature-measuring device must be provided, properly calibrated, and used to ensure proper cooking and holding temperatures. Minimizing the time between preparation and consumption is another effective way to avoid temperature abuse.

Employees must practice good personal hygiene and use proper food-handling practices as described in Chapter 3. They must wear clean clothing and effective hair restraints. They must not eat and smoke around food, and they must wash their hands whenever they are contaminated. Employees shall use tongs, utensils, and deli tissue to avoid bare hand contact with food. Disposable gloves can provide an additional barrier against contamination. However, gloves must not be viewed as a substitute for proper handwashing.

Food must be protected from contamination and cross contamination. Food on display must be protected from contamination by customers and employees. Food equipment must be designed and constructed to make it smooth, easily cleanable, nontoxic and nonabsorbent.

Use disposable utensils whenever possible. When it is necessary to use multiple-use utensils, they must be washed and sanitized using a four-step process:

1. Washing in hot, soapy water.
2. Rinsing in clean water.
3. Chemical sanitizing.
4. Air drying.

A detailed explanation of this process is presented in Chapter 6, *Cleaning and Sanitizing Operations*. It is a good idea to have additional cleaned and sanitized utensils on hand to accommodate peak periods when there may not be adequate time to perform the dishwashing process. Cloths used to wipe up spills should be stored in a pail of sanitizer solution between uses. The concentration of the sanitizer must be regularly checked with test kits or test strips to make certain it remains at the required strength. Always replace the solution whenever it falls below the manufacturer's required minimum concentration.

Garbage and paper wastes should be placed in containers that are lined with plastic bags and equipped with tight-fitting lids. Food and wastes must be kept covered to avoid attracting insects, rodents, and other pests.

Vending Machines

A vending machine is a self-service device which dispenses individual sized servings of food and beverages after a customer inserts a coin, paper currency, token, card, key, or makes a payment by another means. Some vending machines dispense food and beverages in bulk while others dispense products in individually wrapped packages. Vending machines are available that will dispense potentially hazardous, ready-to-eat foods (i.e., sandwiches, french fries, dairy products, and soup) and non-potentially hazardous foods (candy, snack chips, pastries, coffee and soft drinks).

Vending machines storing and dispensing potentially hazardous foods must have adequate refrigeration and/or heating units, insulation, and controls to keep cold foods cold and hot foods hot. These machines must also be equipped with an automatic control that prevents the machine from vending food if there is a power failure, mechanical failure or other condition that prevents cold food from being maintained at 41°F (5°C) or below and hot food maintained at 140°F (60°C) or above.

The temperature cut-off requirement does not apply:

- ▲ In a cold food machine during a period not to exceed 30 minutes immediately after the machine is filled, serviced or restocked; or
- ▲ In a hot food machine during a period not to exceed 120 minutes immediately after the machine is filled, serviced or restocked.

Refrigerated, ready-to-eat, potentially hazardous foods prepared in a retail food establishment and dispensed through a vending machine (with an automatic shutoff control) shall be discarded if it (a) exceeds the time and temperature combinations prescribed by the Food Code or (b) is not correctly date labeled.

Home Meal Replacement

Home meal replacement and meal solutions are the terms most often used to refer to high-quality meals prepared away from home but eaten at home. These types of products are now a multi-billion dollar business for retail food establishments throughout the country. The move toward prepared food has retail food establishments selling complete meals instead of just ingredients. Retail food establishments and restaurants are now competing head-to-head for the "heat and eat" business that has become very popular with today's customers.

Home meal replacements come in "ready-to-cook," "ready-to-heat," and "ready-to-eat" varieties. All three varieties are designed to save time and effort for families that are too tired to cook at the end of the day.

Common examples of "ready-to-cook" foods include beef and chicken kebobs with the peppers and onions already cut up or stuffed green peppers. These foods have found their way into most retail food establishments' meat or frozen food departments.

"Ready-to-heat" food is a much broader segment of the market. It includes the fully cooked hams, turkey roast, or breaded chicken cutlets found in the meat departments of retail food establishments. Many retail food establishments have expanded their deli sections to include items that can be conveniently reheated, such as precooked egg rolls, burritos, chicken dinners, and other foods in the cold food case.

"Ready-to-eat" foods include pizzas, rotisserie chickens, and hot-tray choices such as macaroni and cheese, meat loaf, and mashed potatoes. Customers stop by on the way home from the office and buy complete meals ready to eat right out of the box.

With home meal replacements, retailers must deliver quality, consistency, and safe food and have it ready when it is picked up by or delivered to their customers. By following prescribed food safety guidelines for time and temperature, retailers can assure that meals provided to their customers are safe and wholesome. Hot foods must be cooked to the required product temperature for the required amount of time. These foods must be held at 140°F (60°C) or above while on display and during delivery. Cold foods must be chilled quickly and held at 41°F (5°C) or below during display and delivery.

Home replacement meals should be labeled so customers understand how to keep the product safe when they take it home. Warn them against keeping the food in the car while they do more shopping or keeping it at room temperature when they get home. Pamphlets and brochures stuffed into bags or stapled to the front of bags can help educate customers about safe handling of home replacement meals. In addition, it is a good idea to establish "sell-by" or "best if used by" dates and codes for these foods. Retailers must be able to show they've done everything in their power, including written documentation and following industry standards, to make the food safe.

Summary

The case study presented at the beginning of this chapter shows how bacteria can be transferred from raw animal foods to cooked and ready-to-eat foods (ham and cheese sandwich) via cross contamination. Raw chicken is a common source of *Salmonella* bacteria. Cross contamination can occur in retail food establishments when raw and ready-to-eat foods are handled in the same food production areas. Cross contamination can be prevented by practicing good personal hygiene, especially proper handwashing, proper cleaning and sanitizing of equipment and utensils, and by working with raw and cooked and ready-to-eat foods in different areas of the establishment.

Not all retail food establishment managers will actually purchase food products. However, a knowledge of the rules, regulations, and procedures for receiving and storing food is a must for everyone responsible for food safety.

To ensure all foods are safe for human consumption, purchase foods from inspected sources that follow good manufacturing practices (GMPs) and labeling laws. Schedule deliveries for the off peak times when someone is available to inspect incoming supplies. Receiving personnel must assure food is:

- Free from filth or spoilage
- In sound condition
- At the proper temperature.

Merchandise that is damaged, spoiled, or otherwise unfit for sale or use in a retail food establishment must be properly disposed of, held by the establishment for credit, or returned to the distributor. It is important distressed products not be mistaken for safe and wholesome products and/or cause contamination. These products must be segregated and held in designated areas that will keep them separate from food, equipment, utensils, linens, and single-service and single-use articles.

Foods that are delivered must be moved quickly into designated storage areas. Storage serves as a temporary link between receiving and preparation. Storage facilities must be capable of preserving the safety and wholesomeness of products stored there.

Store foods in their original containers or packages whenever possible. When food is removed from the package or container in which it was delivered, it must then be stored in a clean, covered container that is properly labeled and dated. Proper stock rotation using the FIFO system is necessary to ensure stocked items will be used within a reasonable period of time.

Provide adequate "dry storage" for shelf-stable items such as canned goods, cereals, and spices. Refrigerated storage is required for perishable and potentially hazardous foods such as dairy products, meat, poultry, eggs, fish, seafood, fruits, and vegetables. Store these products out of the temperature danger zone. Freezer storage that will keep foods solidly frozen must be provided for all frozen foods.

Refrigerators and freezers must be equipped with a numerically scaled thermometer to ensure the maintenance of all potentially hazardous food at the required cold temperatures during storage. The thermometer should measure temperature in increments no greater than 2°F (1°C). The sensor of the temperature-measuring device must be located to measure the air temperature in the warmest part of a mechanically refrigerated unit. The temperature-sensing device should also be located to allow easy viewing of the device's temperature display. The temperature of food should be measured to verify holding temperatures are correct.

To prevent the possibility of cross contamination, always store raw or unwashed foods below ready-to-eat foods and foods that will not receive any additional cooking. Do not store food containers on the floor or under any exposed or unprotected sewer or water lines.

Keep the storage areas, including the floors, walls, and ceilings, clean. Make sure the shelving units in these areas are clean and in good repair. Spilled foods should be cleaned up immediately and removed from the area.

Errors that occur during the preparation and service of food are the leading causes of foodborne illness in retail food establishments. To ensure safe food, proper food handling is essential.

Safe food preparation and service include good personal hygiene and taking measures to avoid cross contamination. Employees must know when and how to wash their hands properly. They must also know how to clean and sanitize food-contact surfaces of equipment and utensils properly. Monitoring and controlling food temperatures during preparation are important for destroying microorganisms and preventing them from growing in the food. Keeping foods out of the temperature danger zone as much as possible and cooking them properly are both very important considerations. Foods should only be allowed in the temperature danger zone for a short time during thawing, heating, and cooling activities. The three main contributors to foodborne illness in retail food establishments are:

- ▲ Time and temperature abuse
- ▲ Cross contamination
- ▲ Poor personal health and hygiene practices by food handlers.

These factors must be controlled throughout the flow of food to assure food safety. Preparation and service are critical processes in your establishment because they are the last steps you take before your customers eats the food.

Discussion Questions (Short Answer)

1. Why should buyers purchase foods only from approved sources that comply with all applicable food laws?
2. Discuss the differences between grading and inspection services.
3. What range of temperatures should food temperature-measuring devices be able to measure?
4. Why should cans with swollen ends be rejected and sent back to the supplier?
5. What are typical signs of spoilage for red meats, poultry, and fish?
6. What is the meaning of FIFO?
7. Why should raw products be placed below cooked or ready-to-eat foods during storage?
8. Describe safe procedures for storing cleaning and sanitizing agents and pesticides.
9. Why should products be located at least 6 inches off the floor during storage?

10. Why is it important for a food employee to wash his/her hands?
11. Suggest some ways to decrease the cooling time for hot foods.
12. Identify four ways to safely thaw frozen foods.

Quiz 4 (Multiple Choice)

1. Which of the following is not a rule that should be closely followed when purchasing food?
   a. Foods prepared in a private home may not be used or offered for human consumption in a retail food establishment.
   b. Buyers should only purchase food that is safe, wholesome, and from an approved source.
   c. Avoid the use of commercially raised game animals as meat and poultry items.
   d. Only buy meat and poultry that has been inspected by USDA or state agency officials.
2. The best method for measuring the temperature of frozen food products is by:
   a. Inserting the sensing probe into the center of a frozen food package until the recorded temperature stabilizes.
   b. Inserting the sensing probe between two packages of frozen foods until the recorded temperature stabilizes.
   c. Measuring the ambient temperature of the frozen food compartment of the delivery vehicle.
   d. Looking for signs of freezing and thawing, such as large ice crystals and frozen juices in the bottom of the box.
3. Frozen foods should not be accepted at a retail food establishment if:
   a. They have large ice crystals on the surface.
   b. There are frozen juices on the bottom of the package.
   c. If the temperature is above 32°F (0°C).
   d. All of the above.

4. Which of the following foods should not be rejected upon delivery?
   a. Fresh fish that has dull, sunken eyes, and soft flesh.
   b. Poultry with darkened wing tips and soft flesh.
   c. Canned fruit with small amounts of surface rust on the lid of the can.
   d. Fresh beef products that are delivered at 45°F (7°C).
5. Which of the following statements about fish and seafood is false?
   a. Fish and shellfish are less likely to spoil than red meat and poultry.
   b. Quality in fish and shellfish is measured by smell and appearance.
   c. Fish that may be eaten raw must be commercially frozen prior to consumption.
   d. Molluscan shellfish tags must be kept for 90 days from the date the container is emptied.
6. Which of the following storage practices should prompt a manager to take corrective action?
   a. Products in the dry storage area are being rotated on a first-in, first-out stock basis.
   b. Foods stored in the walk-in freezer are stored on slatted shelves that are 6 inches above the floor.
   c. Raw poultry is stored above potato salad in the walk-in refrigerator.
   d. Cleaning and sanitizing agents and pesticides are stored in a locked and labeled cabinet in the dry food storage area.
7. Pork roasts should be cooked to an internal temperature of at least ______ for 4 minutes to be considered safe.
   a. 140°F (60°C).
   b. 145°F (63°C).
   c. 155°F (68°C).
   d. 165°F (74°C).

8. Which of the following is considered the safest way to thaw potentially hazardous foods?
   a. In the microwave oven.
   b. At room temperature.
   c. In the refrigerator.
   d. Under cool running water at 70°F (21°C).
9. Hot foods should be held at _____ or above and cold foods should be held at _____ or below.
   a. 165°F (74°C); 41°F (5°C).
   b. 165°F (74°C); 32°F (0°C).
   c. 140°F (60°C); 41°F (5°C).
   d. 140°F (60°C); 32°F (0°C).
10. Poultry and stuffed meats should be cooked to an internal temperature of ______ for 15 seconds to be considered safe.
    a. 140°F (60°C).
    b. 145°F (63°C).
    c. 155°F (68°C).
    d. 165°F (74°C).
11. Ground beef meats should be cooked to an internal temperature of ______ for 15 seconds to be considered safe.
    a. 140°F (60°C).
    b. 145°F (63°C).
    c. 155°F (68°C).
    d. 165°F (74°C).
12. All potentially hazardous foods that have been cooked and cooled need to be reheated to an internal temperature of ______ within two hours to be considered safe.
    a. 140°F (5°C).
    b. 145°F (63°C).
    c. 155°F (68°C).
    d. 165°F (74°C).

13. According to the *FDA Food Code*, foods that are cooked and then cooled must be cooled in which of the following ways?
    a. From 140°F (60°C) to 41°F (5°C) in 12 hours.
    b. From 140°F (60°C) to 41°F (5°C) in 8 hours.
    c. From140°F (60°C) to 70°F (21°C) in 2 hours, and from 140°F (60°C) to 41°F (5°C) within 6 hours.
    d. From 140°F (60°C) to 70°F (21°C) in 4 hours, and from 70°F (21°C) to 41°F (5°C) in an additional 2 hours.
14. All foods that are to be held cold must be held at _______ or below.
    a. 41°F (5°C).
    b. 50°F (10°C).
    c. 70°F (21°C).
    d. 0°F (-18°C).
15. All foods that are to be held hot must be held at _______ or above.
    a. 70°F (10°C).
    b. 98°F (37°C).
    c. 120°F (49°C).
    d. 140°F (60°C).

Answers to the multiple choice questions are provided in Appendix A.

References/Suggested Readings

Food and Drug Administration (2001). 2001 *Food Code.* United States Public Health Service. Washington, DC.

Food and Drug Administration (2001). *Fish and Fishery Products Hazards and Control Guide* - Third Edition. United States Public Health Service. Washington, DC.

Longreè, Karla, and G. Armbruster (1996). *Quantity Food Sanitation.* John Wiley and Sons, Inc. New York, NY.

Thayer, David W., et al. (1996). *Radiation Pasteurization of Food.* Council for Agricultural Science and Technology Issue Paper, No. 7. April, 1996.

Suggested Web Sites

Gateway to Government Food Safety Information
www.foodsafety.gov

United States Department of Agriculture (USDA)
www.usda.gov

Centers for Disease Control and Prevention (CDC)
www.cdc.gov

United States Food and Drug Administration (FDA)
www.fda.gov

The American Egg Board
www.aeb.org

America Meat Institute
www.meatami.org

The National Chicken Council
www.eatchicken.com

U.S. Poultry and Egg Association
www.poultryegg.org

USDA Foods Safety and Inspection Service (FSIS)
www.fsis.usda.gov

USDA/FDA Food and Nutrition Information Center
www.nal.usda.gov/fnic/

United States Environmental Protection Agency (EPA)
www.epa.gov

Partnerships for Food Safety Education
www.fightbac.org

The Food Marketing Institute
www.fmi.org

5 Facilities, Equipment, and Utensils

Glass from Broken Light Bulb Creates Problem in Deli

A fluorescent light bulb was burned out in the deli department at Mike's market. The burned out bulb was in a fixture located directly above a work counter being used by deli employees to transfer cold meat salads from bulk containers into shallow containers for display.

A maintenance employee was called to replace the bulb. This employee removed the bulb and the plastic sleeve that covered it from the fixture. When removing the plastic sleeve from the bulb, it shattered and pieces of glass fell down on the food and containers below.

What problems have just occurred at the deli counter? What must be done to protect deli customers from injury?

Learning Objectives

After reading this chapter, you should be able to:

- ▲ Identify ways in which the design and layout of facilities contribute to the efficiency and effectiveness of food production and handling in retail food establishments.
- ▲ Understand the importance of purchasing and properly maintaining equipment and utensils and the influence they have on food safety.
- ▲ Describe how work tasks are conducted in work centers and how the preparation and service of food flows through retail food establishments.
- ▲ Identify the criteria that should be used when determining the need for various types of equipment used in retail food establishments.
- ▲ Understand the basic design and construction requirements that apply to floor- and table-mounted equipment used in retail food establishments.

- Explain the role of third-party certification organizations (NSF International, Underwriters Laboratories, etc.) in the approval of food equipment.
- Recognize the different types of cooking, refrigeration, preparation and dishwashing equipment available for use in retail food establishments.
- Describe how proper installation and maintenance affect the operation of equipment used during the course of food production, holding, display, and handling.
- Explain the role of proper lighting in food production and warewashing areas.
- Explain how proper heating, air conditioning, and ventilation affect food sanitation and employee comfort and productivity in retail food establishments.

Essential Terms

Easily movable

Easy-to-clean

Equipment

Food-contact surface

Kitchenware
- **Tableware**
- **Utensils**
- **Single-use articles**

Non-food-contact surface

Sealed

Smooth

Design, Layout, and Facilities

The design, layout, and facilities provided in a retail food establishment must be consistent with the types of foods being sold there. The equipment used in the various departments must be properly designed and sized to meet the production needs of the area. A layout that works well in one establishment may not necessarily be suitable for another that produces and sells different food items. Most equipment and facilities will be used for several years. Therefore, managers should seek a design that meets current needs but aids future expansion and remodeling when needed to accommodate increased sales of new items that are offered for sale in the store.

The design and layout of storage, production, display, and dishwashing areas should provide an environment in which work may be conducted in a safe, sanitary and efficient manner. Facilities planning and design are often left to architects or engineers who may not be familiar with the needs of the establishment. You and your employees can help architects and engineers by providing practical information about your operation during the design phase. Use a team approach to prepare and review the facility plans. You can frequently avoid unnecessary costs and inconvenience by identifying and preventing problems during the design phase of the project.

Some general areas that are commonly found in retail food establishments are:

- Delivery and receiving
- Storage
- Preparation
- Hot- and cold-holding
- Display
- Handling and service
- Warewashing, cleaning, and sanitizing
- Waste storage and pickup
- Housekeeping
- Toilet facilities.

The activities that are carried out by employees in the various departments are called functions. When planning a retail food establishment, try to understand and visualize each function that will be conducted within the different departments. Next, determine the specific tasks employees must complete when performing the function. Finally, the tasks must be arranged in a way that allows a smooth and sequential flow within that department or area. It is important you develop a flow diagram of your establishment's operation in order to plan the physical facilities for each function.

A properly designed and constructed establishment offers many benefits. Attention to the design and construction of the building, the equipment, and its installation ensures high standards of sanitation and safety and helps make maintenance activities easier.

Regulatory Considerations

When planning facilities for retail food establishments, you must know about and comply with national, state, and local standards and codes related to:

- ▲ Health
- ▲ Safety
- ▲ Building
- ▲ Fire
- ▲ Zoning
- ▲ Environmental code standards.

The **equipment** used in retail food establishments should meet the American National Standards Institute (ANSI) standards and bear the stamp of approval of recognized third-party certification organizations (Figure 5.1). Some examples of these organizations include:

- ▲ NSF International
- ▲ Underwriters Laboratories, Inc. (UL)
- ▲ American Gas Association (AGA).

Buyers who purchase equipment approved by these organizations are assured quality materials are used in the construction. Also, the items are designed and constructed to meet accepted food sanitation criteria. A partial list of NSF International Standards is provided at the end of this chapter. Installation of equipment must conform to local code. Officials from local health, building, and fire protection agencies usually conduct on-site inspections of equipment installation.

Figure 5.1 UL and NSF Seals

Work Center Planning

The perishable departments in a retail food establishment are commonly organized into **work centers**. These are areas where a group of closely related tasks are performed by an individual or individuals.

The number of work centers required in a department depends on the number of functions to be performed and the volume of material to be handled.

Two of the most important features to consider when planning work centers are the total space needed and the arrangement of equipment in that space. An employee should be able to complete the related tasks at the work center without moving away from it. The work center should also be large enough to do the job yet small enough to reduce travel and conserve time and effort. A properly designed work center will provide adequate facilities and space for:

- Efficient production
- Fast handling and service
- A pleasant environment
- Effective cleanup.

Equipment Selection

It is extremely important to select the right piece of equipment for the job. Compare different pieces of equipment for a particular job and look at such features as:

- Design
- Construction
- Durability
- Ability to clean easily
- Size
- Cost
- Safety
- Overall ability to do the job.

Need

The basic needs of a retail food establishment should dictate the purchase of equipment. Need should be based on whether the addition of a particular piece of equipment will improve the quality of food, reduce labor and material costs, improve sanitation, and contribute to the bottom line of the establishment.

Establishment managers and their department heads should assign a priority to equipment needs. When buying equipment, look for products that meet current and future needs and demands of the operation, do not require extraordinary repair and upkeep, and will work properly for many years.

Cost

Some of the major costs associated with the purchase of equipment are the purchase price, installation cost, operating costs, maintenance costs, and finance charges.

Compare these costs for the different pieces of equipment you are considering before you make a purchase. Some of these costs will be calculated by the manufacturers and provided for comparison purposes. For others, the establishment manager or his agent will have to collect information to assist in making a purchasing decision. It is not advisable to buy equipment that is more expensive, larger, or more sophisticated than that which is required to meet the needs of the operation.

Size and Design

Equipment should easily fit into the space available in the layout of the facility. You should consider future needs when selecting equipment. However, it is unwise to buy excessively oversized equipment in anticipation of growth many years into the future. Equipment that is too large can cause as much wasted effort as equipment that is too small.

Design is also an important feature of food equipment, since the equipment used in a retail food establishment is subject to constant use and abuse. Equipment and utensils must be designed to function properly when used for their intended purposes.

Equipment that sits on the floor must be elevated on legs, sealed to the floor, or mounted on casters to make it easily movable. **When legs are used, they should provide a minimum clearance of 6 inches (15 centimeters) between the floor and the bottom surface of the equipment** (Figure 5.2). This space makes it easier to clean the floor under and behind the equipment and discourages pest harborage.

Figure 5.2 Floor-mounted Equipment

Heavy equipment such as ranges and coolers may be mounted on a raised masonry, tile, or metal platform and sealed to the floor. This helps prevent trash and pests from collecting under the equipment.

In retail food establishments, display shelving units, display refrigeration units, and display freezer units located in the customer shopping areas do not have to be elevated or sealed to the floor. However, the floor under these units must be maintained in a clean and sanitary manner.

Table-mounted equipment (that is not easily movable) should be on 4-inch (10-centimeter) legs. This provides clearance between the counter top and the bottom of the equipment and makes it easier to clean under and around the equipment. The equipment may also be sealed to the counter.

Construction Materials

Food equipment has food-contact and nonfood-contact surfaces. **Food-contact surfaces** are the parts of equipment and utensils which normally come into contact with food or from which food may drain, drip, splash, or spill into food or onto a surface that is normally in contact with food. **Nonfood-contact surfaces** are the remaining parts of the equipment and utensils and the surrounding area that should not make contact with food during production.

The *FDA Food Code* and construction standards, such as those from ANSI require food equipment and utensils to be:

- ▲ Smooth
- ▲ Seamless
- ▲ Easily cleanable
- ▲ Easy to take apart
- ▲ Easy to reassemble
- ▲ Equipped with rounded corners and edges.

Materials used in the construction of utensils and food-contact surfaces of equipment must be nontoxic and not impart colors, odors, or tastes to foods. Under normal use, these materials must also be safe, durable, corrosion-resistant, and resistant to chipping, pitting, and deterioration.

Metals

Metals are very popular materials in retail food establishments. Chromium over steel gives an easily cleanable, high-luster finish. It is commonly used in conjunction with small appliances. Noncorrosive metals formed by the alloys of iron, nickel, and chromium may also be used in the construction of food equipment.

Lead, brass, copper, cadmium, and galvanized metal can cause a chemical poisoning when they come into contact with high acid foods (foods that have a low pH). Therefore, these materials must not be used as food-contact surfaces for equipment, utensils, and containers.

Stainless Steel

Stainless steel is one of the most popular materials in retail food establishments. It is commonly the material of choice for food containers, table tops, sinks, dish tables, dishwashers and ventilation hood systems.

Stainless steel has a durable, shiny surface that easily shows soil and is **easy-to-clean** and maintain. Stainless steel also resists high temperatures, rust, and stain formation.

Do not use abrasive cleaners and scouring pads to clean stainless steel. These cleaning supplies can scratch the surface of the metal and cause harborage areas for disease-causing microorganisms.

One factor that influences the cost of stainless steel is the extent of polishing desired, because polishing requires labor, material, and energy. Finish No. 4 is most commonly preferred for food production areas whereas higher finishes are usually preferred in display and handling areas. The polish of the stainless steel is indicated by numbers 1-8. The higher the number, the higher the polish.

The gauge of the stainless steel is defined as the thickness of the product. The most common gauges are 14, 16, 18, 20, and 22. The higher the number of gauge, the thinner the steel. The gauge you use should be determined by the particular use. For example, 14 and 16 gauge is ideal for tops of worktables, sinks and dish tables while 20 is ideal for cabinet doors.

Plastic

Plastic and fiberglass are very popular in retail food establishments because they are durable, inexpensive, and can be molded into different combinations. Undoubtedly, various types of plastic will be used more and more as components of food equipment in the future.

Some examples of plastics used in retail food establishments are:

- ▲ Acrylics (used to make covers for food containers)
- ▲ Fiberglass (used in boxes, bus trays, and trays)
- ▲ Polyethylene (used in storage containers and bowls).

You must be sure to buy food equipment that is made of only food-grade plastics. There are many different brand names for plastics. Select the one that works best for you based on intended use and durability. The harder, more durable plastics are easier to clean and sanitize. Consult NSF Standard No. 51 when selecting plastic materials and coverings.

Wood

Wood has both advantages and disadvantages when used as food-contact surfaces. The primary advantages are it is light in weight and economical. The disadvantages are the fact wood is porous to bacteria and moisture, and it absorbs food odors and stains. Wood also wears easily under normal use, which requires frequent maintenance and replacement (See Figure 5.3).

The *FDA Food Code* permits limited use of wood materials including hard maple or an equally hard, close-grained wood for cutting boards, cutting blocks, and baker's tables. Wood is also approved for paddles used in pizza operations.

The consensus seems to be the disadvantages of using wood for food-contact surfaces outweigh the advantages. This is primarily because of the high degree of maintenance required and problems with keeping the equipment clean and sanitary.

Figure 5.3 Wood vs. Plastic Materials

Types of Equipment

In order to select the proper equipment for a retail food establishment, it is necessary to obtain up-to-date information about the equipment from manufacturers' brochures and catalogues, and trade journals and magazines.

NSF International and Underwriters Laboratories, Inc., provide independent evaluations of equipment and materials. The NSF and UL emblems on a piece of equipment verify the equipment has been tested and meets the requirements prescribed in the ANSI standards. Sample pieces of the equipment are commonly tested for performance in the laboratory and in the field. For additional information refer to the reference/suggested reading section at the end of this chapter.

Cooking Equipment

The most important criteria to use when selecting cooking equipment are the types and quantities of food prepared, ease of cleaning, durability and energy conservation. The frame, door, exterior, and interior materials of cooking equipment should contribute to the durability and cleanability of the equipment. The type and thickness of insulation materials contribute to the energy efficiency of this equipment.

Ovens

Ovens are important pieces of equipment in the bakery and may well be used by other departments in a retail food establishment. Ovens are used to bake a variety of foods to different temperatures. The heat in an oven is distributed by radiation, conduction, or convection, depending on the type of oven being used.

A good oven should rise to 450°F (232°C) within 20 minutes, and proper heat circulation is important. Ovens should be able to cool quickly when a drop in temperature is required.

The ***range oven*** is the most commonly used oven in small operations (Figure 5.4). It has a cooking surface above the oven that makes it suitable for doing all kinds of work.

Figure 5.4 Range Oven

(Courtesy of Hobart Corp.)

Range ovens must have good insulation and construction to prevent heat loss. They should also be equipped with heavy-duty doors and shelves. A window in the oven door makes it easier to see inside.

Rotisserie ovens are commonly used in retail food establishments to cook chickens. While chicken is the most popular rotisserie product, these ovens can also be configured to accommodate turkeys, beef roasts, and ducks. Rotisserie ovens cook foods with gas or electricity and some models are designed to use woods such as mesquite or hickory. Some rotisserie ovens have dual cooking compartments that work independently. By adjusting thermostatic controls, an operator may use both sections to roast, one to warm and one to roast, or both sections to keep products warm.

The cooking and warming compartments of rotisserie ovens must be designed and constructed to promote easy cleaning. Electrical units do not normally require ventilation. However, since the surrounding environment may get hot, a hood may be useful. Gas units must be vented, either under a hood or directly vented via a flue.

Figure 5.5 Rotisserie Oven

Deck ovens are the stationary type where one deck sits on top of another to conserve space (Figure 5.6). This type equipment is commonly used to bake pizza, cookies, and muffins.

Figure 5.6 Deck Oven
(Courtesy of Hobart Corp.)

The heated air in a deck oven is circulated around the outside of the heating chamber and radiates through the lining. Deck ovens contain separate heating elements and each unit is controlled separately. Good insulation must be provided between the decks. The standard deck oven is designed to provide different working heights with 7 to 8 inches clearance provided for baking and 12 to 16 inches clearance for roasting.

A ***convection oven*** uses a high-speed fan to circulate heated air and guarantee even heat distribution over and around foods (Figure 5.7). Cooking time in a convection oven is about one-third to three-fourths the amount required in deck ovens. Another advantage of the convection oven is more cooking is accomplished in a smaller space, as food is placed on multiple racks instead of on a single deck.

Figure 5.7 Convection Oven

(Courtesy of Hobart Corp.)

Microwave ovens are commonly used for thawing, heating, and reheating food (Figure 5.8). Microwaves are electromagnetic waves of radiant energy which, like light, travel in straight lines. Microwaves are reflected by metals, pass through air, and are absorbed by several food components. They pass through many, but not all, types of paper, glass, and plastic materials. When microwaves are reflected or pass through a material, they do not give off heat to the object. To the extent that microwaves are absorbed, they heat the absorbing material.

Figure 5.8 Microwave Oven

The advantage of the microwave oven is it can cook small quantities of food very rapidly. A microwave oven is not useful for cooking large quantities of food due to its limited capacity. However, it is excellent as a support to quantity cooking.

Other types of ovens may be used in retail food establishments. These include rotary, infrared, conveyor, and roll-in units. Each of these pieces of equipment has unique features and has been designed for special applications. You should contact your equipment supplier to determine if this type of equipment is best suited to carry out the functions required in your operation.

Refrigeration and Low-Temperature Storage Equipment

Refrigeration is an important feature of the safe transport and storage of perishable foods. Cold-holding permits us to have an ample supply of meats, poultry, fish, dairy products, fruits, and vegetables throughout the country during practically all seasons of the year. Refrigeration during transportation, combined with improved cold storage facilities, has helped to stabilize the price of perishable foods.

Refrigerators and freezers are used to protect perishable foods and preserve the safety and wholesomeness of potentially hazardous foods. Many foods deteriorate rapidly at room temperature. Spoilage, waste, and shrinkage can be greatly reduced by keeping foods at lower temperatures until they are used.

Proper cooling requires removing heat from food quickly enough to prevent microbial growth. Improper cooling of potentially hazardous foods is consistently identified as one of the leading contributors to foodborne illness. Bacteria grow best at temperatures between 70°F (21°C) and 120°F (49°C). **The *FDA Food Code* states cooked potentially hazardous foods must be cooled from 140°F (60°C) to 70°F (21°C) within two hours and from 140°F (60°C) to 41°F (5°C) or below within six hours.**

All types of refrigerators and freezers have a maximum capacity for cooling foods. Too often, employees put large amounts of hot food in a unit and rely on it to "cool" the food. On the contrary, the addition of large amounts of hot food causes the inside temperature of the unit to rise above acceptable storage temperatures. This, in turn, causes food in the unit to be stored in the temperature danger zone. Don't forget the primary purpose of a refrigerator is to take foods that are already at or near the proper cold-holding temperature and keep them out of the temperature danger zone.

The efficient operation of a refrigeration unit depends on several factors including:

- ▲ Design
- ▲ Construction
- ▲ Capacity of the equipment.

Proper air circulation both inside and outside the refrigerator is important. **The storage of food in shallow containers, placed on slatted shelves or tray slides to permit good circulation of the chilled air**, is essential for both short- and long-term storage.

Do not line shelves in refrigerators and freezers with sheet pans, foil, plastic, or cardboard. This decreases air flow in the storage compartment and reduces the cooling capacity of the equipment. Proper construction of cold storage units includes sturdy construction of doors, hardware, and fixtures. Doors may be full- or half-length, and shelving should be adjustable. Door gaskets should be durable and easy-to-clean. They must be replaced when worn.

The size of refrigerator or freezer needed depends on the size of the work area and the type and amount of food to be stored in the unit. Refrigerators and freezer units have far more capacity today than ever before. Refrigeration and low-temperature storage equipment must be adequately sized and properly installed to assure reliable and efficient operations (Figure 5.9).

Figure 5.9 Refrigeration Equipment

Maintenance of the refrigeration equipment in the various departments is an important responsibility. These units must be cleaned on a regular basis to maintain good sanitary conditions and eliminate odors. The inside walls, floor, shelves and other accessories of a walk-in refrigeration unit must be cleaned regularly to remove spills and debris. Don't forget to clean the fan grates and condenser as part of your routine cleaning.

Instruct employees to open the doors of refrigerators as briefly and infrequently as possible when taking food in and out of storage and display. Large amounts of warm, moist air can raise the holding temperature and humidity inside a refrigerator unit very quickly.

Reach-in Refrigeration

Several types of reach-in refrigeration units can be used for cold storage in retail food establishments. Some of the more popular models are *upright, under-the-counter,* and *mobile* units. Reach-in refrigerators are commonly found in the deli, bakery and other departments of a retail food establishment.

The capacity of reach-in refrigerators varies greatly. Upright models can have a capacity of nearly 90 cubic feet. These models are usually equipped with single, double, or multiple door configurations [Figure 5.10(a)]. Under-the-counter refrigeration units typically have only a few cubic feet of storage space [Figure 5.10(b)]. These units can be placed below food preparation counters and provide on-the-spot storage.

a. Standard Upright b. Under-the-Counter

Figure 5.10 Reach-in Refrigerators

(Courtesy of Hobart Corp.)

Refrigerated Display Cases

Retail food establishments commonly use coffin, upright/vertical, and reach-in refrigerated cases to hold cold and frozen foods during display (Figure 5.11).

a. Coffin b. Upright/vertical c. Reach-in

Figure 5.11 Refrigerated Display Cases

This equipment is designed to keep cold foods out of the temperature danger zone and frozen foods solidly frozen while on display. Employees must not exceed the cooling capacity of these merchandisers by stocking cold and frozen foods above the maximum load line. Employees must also avoid covering the ventilation discharge and return air openings in upright/vertical refrigerated display cases with dairy, deli and other types of products. The defrost cycles in these merchandisers must be controlled to keep food items from being held in the temperature danger zone for long periods of time.

Walk-in Refrigerators

Walk-in refrigerators are used to store large quantities of perishable and potentially hazardous food at temperatures between 32°F (0°C)and 41°F (5°C). They can also be used to safely thaw products removed from a freezer. In some retail food establishments, walk-in refrigerators are combined with reach-in dairy/deli display cases. The units are loaded from the back which promotes the FIFO rotation of products.

Walk-in refrigerators should be located at a site easy to get to from receiving, production and handling areas. The floors of walk-ins should be made of strong, durable, and easily cleanable material. The walls should have a washable finish that will resist the effects of moisture. A safety device for opening the door should be provided on the inside of the unit.

Some walk-in units have the door opening covered with 4-inch wide plastic strips called a strip curtain. These curtains reduce air loss and conserve energy when the door is opened. An outside wall-mounted thermometer or recorder will enable you to determine the air temperature inside the walk-in cooler without opening the door. The sensing part of the thermometer must be placed in the warmest part of the cooler.

Cook-Chill and Rapid-Chill Systems

Cook-chill is a system in which food is:

- ▲ Cooked using conventional cooking methods
- ▲ Rapidly chilled using a chiller
- ▲ Stored for a limited time
- ▲ Reheated before service to the customer.

In a standard cook-chill system, the food is cooked and then placed in sealed plastic pouches. The pouches are placed in a chill tumbler which looks similar to a front loading clothes washing machine. The pouches are then tumbled inside the machine in ice cold water. The food is typically chilled to 37°F (4°C) in 90 minutes or less and is stored at temperatures between 33°F (1°C) and 38°F (4°C) for five days. Day 1 is counted as the day of production and Day 5 is the day of service.

The advantages of this system include reduction of peaks and valleys in production and readily available foods. The disadvantages of this system include the high cost of the chiller and the storage space requirements.

Rapid-chill systems are designed to cool hot foods very quickly (see Figure 5.12). This type equipment can typically get a few hundred pounds of hot food through the temperature danger zone in two hours or less. Although this equipment is somewhat expensive, it can be a very good investment for some retail food establishments.

The packaging materials for products prepared using a cook-chill system are usually plastic pouches fastened with a metal clip. This clip is a physical hazard you need to keep an eye on. Be sure to remove the metal clip carefully and discard it immediately after the package is opened.

From time to time, retail food establishments may experience a power failure. Without alternate power sources, this may lead to a loss of refrigeration. It is important you have an action plan in place to deal with power failures and keep food safe.

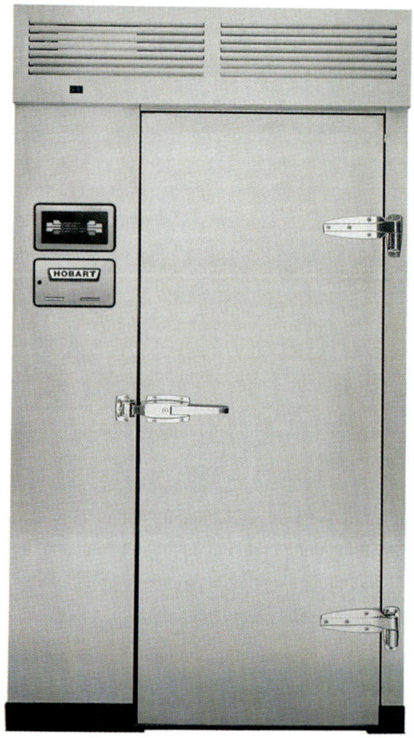

Figure 5.12 Rapid-Chill Equipment
(Courtesy of Hobart Corp.)

First, you should do everything you can to keep the food inside refrigerators cold. Keep the refrigerator door closed as much as possible to keep the cold air inside the unit.

Depending on the type of refrigeration unit, food may be kept out of the temperature danger zone even when the power is off. Avoid opening the doors on refrigeration equipment during a power failure. During a power outage, you must monitor the time and temperature to make sure potentially hazardous foods have not been stored in the temperature danger zone for more than four hours.

Hot-Holding Equipment

Potentially hazardous food that has been cooked and is to be eaten hot must be held at 140°F (60°C) or above. An important fact to remember is hot-holding equipment will not raise the temperature of foods very much. In order for hot-holding equipment to work properly, the food must be at 140°F (60°C) or above when it is put into or onto this equipment.

Hot-holding equipment uses steam, heating elements, or light bulbs to keep foods hot. Some examples of hot-holding equipment that may be found in retail food establishments are shown in Figure 5.13.

Figure 5.13 Hot-Holding Equipment

Hot-holding equipment should be checked regularly to make sure it is working properly. Food temperatures must be monitored to assure they are being maintained at 140°F (60°C) or above. Hot-holding units must be cleaned regularly, and food should be rotated using a FIFO procedure.

Other Types of Food Equipment

Slicers

The basic design of food slicers includes a circular knife blade and carriage that passes under the blade (Figure 5.14). Foods to be sliced are placed on the carriage and fed either automatically or by hand. Thickness of the slices can be controlled by changing the setting of a close adjustment to the blade.

Slicers can be dangerous if not used properly. The manufacturer's instructions should always be followed, and employees should receive training on safe operation and proper cleaning of this equipment.

Figure 5.14 Slicer
(Courtesy of Hobart Corp.)

Mixers, Grinders, Choppers, Tenderizers and Saws

Mixers, grinders, choppers, tenderizers, and saws are examples of equipment commonly found in a retail food establishment. Some of these pieces of equipment are available as table and floor models.

Mixers can also be used to shred and grind when different accessories and attachments are used. Floor model mixers have three standard attachments: (1) a paddle beater for general mixing which can be used to mash, mix, or blend foods and ingredients; (2) a whip to incorporate air into products; and (3) a dough hook used to mix and knead dough.

Mechanical tenderizers are commonly used to improve the edible quality of meat. Tenderizers cut connective tissue in meats and poultry by making incisions with stainless steel blades. The amount of tenderizing is controlled by the depth and the spacing of the blades and the frequency with which the meat is penetrated.

Grinders and choppers (Figure 5.15) work well with a variety of fresh foods and ingredients including meats and vegetables.

Figure 5.15 Meat Grinder

(Courtesy of Hobart Corp.)

Band saws are commonly used in meat and fish processing areas. Heavy-duty saws are best suited for large volume retail food establishments (Figure 5.16). These saws are designed to cut meat and bone and hold up under continuous heavy use. Table top and light-duty saws are commonly used in small delis, meat shops, and small establishments. This equipment is excellent when cutting boneless meats, chicken, and fish.

Band saws must be durable, easily cleanable and safe to use. Food-contact surfaces must be made of nontoxic materials that provide good sanitation qualities. Employees must be able to disassemble this equipment for easy cleaning, preferably without tools. Band saws must be properly guarded or have a "dead man" switch to protect the operator from contact with the blade.

Figure 5.16 Band Saw
(Courtesy of Hobart Corp.)

Ice Machines

Ice is an important item in retail food establishments. It is used to chill beverages and preserve the freshness of fish and some produce items during display.

Ice is food and must be handled with the same degree of care as other food items. Ice must be made from potable water, and ice machines must protect the ice during production and storage (Figure 5.17).

The parts of an ice machine that come into contact with the ice must be smooth, durable, easily cleanable and constructed of nontoxic materials. These food-contact surfaces must be cleaned and sanitized regularly to prevent the growth of mold and other microorganisms. The drain line from the ice machine must be equipped with an air gap to protect the ice from contamination due to backflow. You will learn more about backflow and air gaps in Chapter 7.

Figure 5.17 Ice Machine

Scoops, shovels, carts, and other equipment used to dispense or transport ice must meet the design and construction criteria for food-contact surfaces. When not in use, this equipment should be stored in a manner that will protect it from contamination.

When scoops are stored inside an ice machine, they must be placed in a bracket that is mounted to a wall of the ice storage compartment. This will permit employees to grab the handle of the scoop without their hands touching the ice or the food-contact surface of the scoop.

Employees must never dispense ice by passing a glass or cup through the ice. This can cause glass from the container to break off and become a physical hazard in the ice. Food and beverage containers must not be stored in ice that will be used for drinking purposes (see Figure 5.18). This prevents contamination of the ice your customers may consume.

Figure 5.18 Do Not Store Food and Beverage Containers in Ice Served to Customers

Ambient Temperature Display Equipment

Many foods in retail food establishments are displayed at room temperature. These include shelf-stable products, certain types of produce, bakery items that have a low water activity (bagels, cookies, doughnuts, etc.) and bulk foods such as candy, peanuts, and cereals. These are not considered potentially hazardous foods, and they can be held safely at room temperature. However, they must be protected from contamination by customers and the general environment.

Ambient display equipment can have both food-contact and non-food-contact surfaces. The food-contact surfaces of the equipment must be cleaned and sanitized regularly to remove contaminants. Non-food-contact surfaces must be cleaned regularly to remove spills and eliminate soil.

Stock on display should be rotated using the FIFO method, and utensils should be provided so customers will not contaminate unpackaged food with their bare hands.

Some produce display equipment is equipped with misters that help assure the freshness of fruits and vegetables and prevent wilting, dehydration and shrink (Figure 5.19). These misters must be connected to a potable water supply, and the system and heads of the system must be properly cleaned and maintained to assure maximum freshness, safety, and wholesomeness of produce that is exposed to the mist.

Figure 5.19 Produce Case with Misting Heads

Live Seafood Display and Holding Tanks

Fish, crustacean shellfish (crabs and lobsters), and molluscan shellfish (oysters, clams, and mussels in the shell) are live and perishable products that must be protected to remain safe, wholesome, and attractive to customers (Figure 5.20). The quality of the product in live seafood tanks is only as good as the water in the tank. Water quality is critical, and the key to maintaining optimum water quality is filtration.

Some important handling and maintenance steps to use with this type of equipment are:

- ▲ Clean and replace the ultraviolet light to assure proper intensity to kill pathogens
- ▲ Cull out and discard dead, cracked and weak molluscan shellfish daily
- ▲ Before adding molluscan shellfish to a tank, make sure they are cleaned thoroughly and cull out dead, cracked and weak shellfish
- ▲ Never mix molluscan shellfish with crustacean shellfish

- Never allow molluscan shellfish to be submerged in the system
- Never cook dead or dying shellfish from live displays as a way to reduce shrink
- Clean the tank interior to keep it free of algae and prevent slime buildup which may harbor bacterial pathogens.

Figure 5.20 Live Seafood Display System

The product in a lobster or fish tank relies on the staff of the retail food establishment to keep its environment healthy.

Single-Service and Single-Use Articles

Single-service articles include tableware, carry-out utensils, and other items such as bags, containers, stirrers, straws and wrappers. They are designed and constructed to be used only one time by only one person. After one use the article is discarded.

Single-use articles include items such as wax paper, butcher paper, plastic wrap, and certain types of food containers that are designed to be used once and discarded.

A retail food establishment should provide single-use and single-service articles for food handlers if it does not have proper facilities for cleaning and sanitizing multi-use kitchenware and tableware (see Chapter 7). These establishments must also provide single-service articles for use by consumers.

Materials used to make single-service and single-use articles must not permit the transfer of harmful substances or pass on colors, odors, or tastes to food. These materials must be safe and clean when used in retail food establishments.

Warewashing Equipment

Warewashing is the process used to clean and sanitize the equipment, utensils, dishes, glasses, etc., that are used during the preparation, handling and consumption of foods. Proper warewashing is one of the most important jobs in a retail food establishment. As you learned in Chapter 3, contaminated equipment and utensils have been identified as sources of contamination and cross contamination that can cause foodborne illness.

Equipment and utensil warewashing should be performed in a room or area that is separate from food production areas. In small establishments, there may be only one warewashing area for the entire store. In larger establishments, there may be a warewashing room or area for each department in the facility.

The design of a warewashing area is determined by the volume of equipment and utensils to be washed and the time required to complete the job. As the volume of your business grows, you should evaluate the warewashing operation to make certain it will effectively handle the needs of your operation. In all cases the warewashing area must be well lighted and well ventilated. Noise-absorbing materials may be installed on walls and the ceiling to lower noise levels in warewashing areas.

Some of the items most frequently washed and sanitized in a retail food establishment are utensils such as knives, forks, spoons and tongs, and **kitchenware** such as pots, pans, cutting boards, slicers, grinders, and mixers. Establishments that sell food customers consume onsite may also wash **tableware** such as dishes, glasses, and eating utensils.

The purpose of warewashing is to clean and sanitize equipment and utensils. It consists of two phases:

- A cleaning phase where visible soil is removed from the surface of the item through washing and rinsing
- A sanitizing phase where the number of disease-causing microorganisms on a cleaned surface is reduced to safe levels.

Cleaning and sanitizing operations can be performed either manually or mechanically. Most retail food establishments use manual warewashing operations to clean and sanitize equipment and utensils.

Some larger establishments may use a mechanical dishwashing machine to perform these tasks.

Manual Warewashing

A manual warewashing area must provide adequate space to store soiled equipment and utensils. Items to be cleaned must be pre-flushed or pre-scraped and, if necessary, pre-soaked to remove food particles and soil. A hose and nozzle or other device must be provided to pre-flush and pre-scrape food soil into a garbage container or disposal. This equipment must be located at the soiled end of the warewashing operation to avoid contaminating cleaned and sanitized equipment and utensils.

Most manual warewashing operations are performed in a three-compartment sink (Figure 5.21). In the first compartment, pre-scraped and pre-flushed equipment and utensils are washed in a detergent solution with hot water. Soap and soil are rinsed off equipment and utensils in the second compartment. The third compartment of the sink is used to sanitize the previously cleaned surfaces. Chemicals or hot water may be used to sanitize equipment and utensils in retail food establishments.

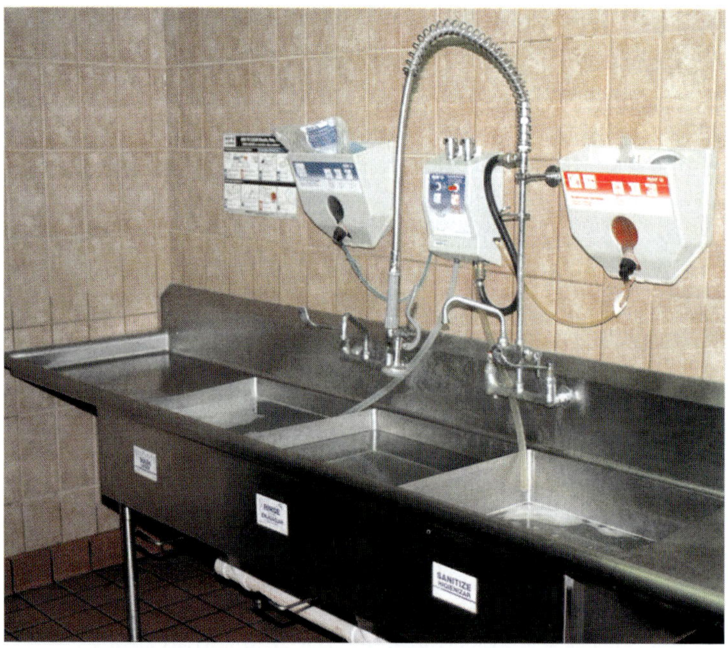

Figure 5.21 Three-Compartment Sink

Manual warewashing can also be performed using a bucket and brush, wall-mounted hose units, and spray units. These processes will be explained in more detail in Chapter 6 of this book.

The compartments of the three-compartment sink must be large enough to accommodate the largest pieces of equipment and utensils used in the food establishment. Supply each compartment with hot and cold potable running water. Provide drainboards or easily movable dish tables of adequate size for proper handling of soiled utensils prior to washing, and for air drying cleaned and sanitized items.

If hot water is used to sanitize equipment and utensils in a manual warewashing operation, the temperature of hot water must be maintained at or above 171°F (77°C). Achieving this amount of heat requires a booster heater or an electrical heating element that can be immersed directly into the water in the third compartment of the sink. Employees must be instructed to use dish baskets or racks to lower equipment and utensils into the very hot sanitizing water. This will prevent employees from being scalded when immersing equipment and utensils in the very hot water.

Mechanical Warewashing

Mechanical warewashing machines may be used by retail food establishments but are far less common than a three-compartment sink. Dishwashing machines can be used to clean and sanitize equipment and utensils that do not have electrical parts and that will fit into the machine. The most common types of mechanical dishwashers used in retail food establishments are single-tank, stationary-rack dishwashers and low-temperature dishwashing machines.

When using a single-tank, stationary-rack dishwashing machine, the dishes are placed on racks and washed one rack at a time with jets of water within a single tank. They are operated by opening a door, inserting a rack of dishes, closing the door, and starting the machine (Figure 5.22).

A dishwashing machine must automatically dispense detergents and sanitizers. These machines must have a visual means to verify detergents and sanitizers are delivered or a visual or audible alarm to signal if the detergent and sanitizers are not delivered to the respective washing and sanitizing cycles.

Low-temperature dishwashers are similar in design to the single-tank, stationary-rack dishwashing machine. However, they use chemicals to sanitize equipment and utensils. This allows lower water temperatures, which conserves energy.

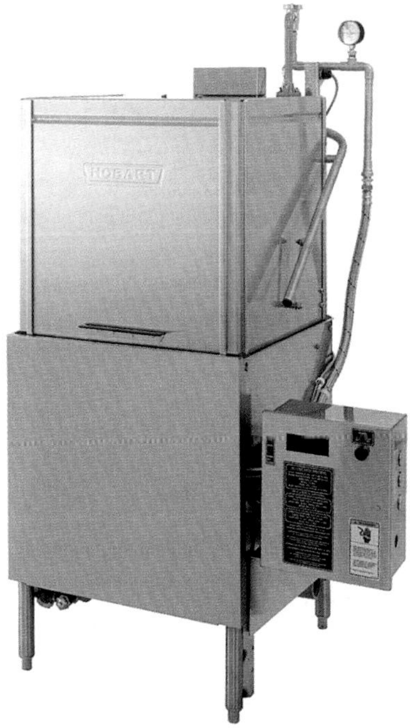

Figure 5.22 Single-Tank Dishwashing Machine
(Courtesy of Hobart Corp.)

The size and design of mechanical dishwashing facilities must meet the needs of the operation. The machine must be large enough to accommodate the size and volume of equipment and utensils to be cleaned and sanitized. Manufacturers can provide specifications for the different types of equipment they make.

When purchasing a dishwashing machine, you should also consider the cost of operation and maintenance. An adequate supply of very hot water is required for the final rinse in a high-temperature dishwashing machine. The high temperatures required for effective sanitization make it expensive and unsafe to maintain general purpose hot water at these temperatures. Therefore, a separate booster heater is needed to raise the temperature of the sanitizing rinse water to the proper temperature. The booster heater must be properly sized, installed, and operated to deliver water at the volume, flow pressure, and temperature required for the operation. For more information about booster heaters, consult NSF International Standard No. 5, Commercial Hot Water Generating Equipment.

Installation

Proper installation is required to assure equipment functions properly. The best design and construction will be worthless if electrical, gas, water, or drain connections are inadequate or improperly installed. The dealer who sells you equipment may or may not be responsible for its installation. Arrangements for installation will usually be specified in your purchase agreement.

Architects, engineers, and contractors are responsible for providing:

- ▲ Adequate plumbing
- ▲ Electrical wiring
- ▲ Venting facilities for the satisfactory installation of equipment
- ▲ Compliance with the standards of local building, plumbing, electrical, health, and fire safety codes.

The operation of a piece of equipment should be tested by engineers, architects, and contractors before it is put into routine service. Employees must also be trained to operate equipment correctly and safely.

Maintenance and Replacement

The cost of care and upkeep on a piece of equipment may determine whether or not its purchase and use are justified. Successful maintenance of equipment requires definite plans to prolong its life and maintain its usefulness. Such plans place emphasis on a few simple procedures:

- ▲ Keep the equipment clean
- ▲ Follow the manufacturer's printed directions for care and operation
- ▲ Post the instruction card for a piece of equipment near it
- ▲ Stress careful operation and maintenance as essential to continued use
- ▲ Make needed repairs promptly.

Some valuable suggestions for the care of equipment are:

- ▲ Assign the care of a machine to a responsible person
- ▲ Check the cleanliness of machines daily
- ▲ Have repairs performed promptly and by a properly trained person.

Lubricants used on food equipment may directly or indirectly end up in food. Therefore, all lubricants used on food-contact surfaces, bearings and gears, and other components of equipment that are located so the lubricants may leak, drip or be forced into food or onto food-contact surfaces must be approved as food additives or generally recognized as safe. These lubricants should be used in the smallest amount needed to get the job done.

Lighting

Good lighting is a very important feature of a work area. Proper lighting in production and dishwashing areas increases productivity, improves workmanship, reduces eye fatigue and employee irritability, and decreases accidents and waste due to employee error. Proper lighting also shows soil and when a surface has been properly cleaned.

Food production and warewashing areas should be furnished with the proper amount of lighting and soft colors that reduce glare. Locate light to eliminate shadows on work surfaces or glare and excessive brightness in the field of vision. Glare causes eye fatigue, reduces efficiency, and increases accidents.

The amount of light required in a department or area depends on the kind of work performed there. The following light intensity levels are recommended in the *FDA Food Code*:

▲ 10 *foot-candles at a distance of 30 inches above the floor*
  ▼ Walk-in refrigeration units
  ▼ Dry food storage areas
  ▼ Other areas and rooms during periods of cleaning.

▲ 20 *foot-candles at a distance of 30 inches above the floor*
  ▼ Areas where fresh produce and packaged foods are offered for sale and consumption
  ▼ Areas used for handwashing
  ▼ Areas used for dishwashing
  ▼ Areas used for equipment and utensil storage
  ▼ Toilet rooms.

▲ 50 *foot-candles at the work surface*
  ▼ Where an employee is working with unpackaged, potentially hazardous food
  ▼ Where an employee is working with food, utensils, and equipment such as knives, slicers, grinders, or saws

- ▼ Where employee safety is a factor.

The food regulations in some state and local jurisdictions may require higher illumination levels in retail food establishments, especially during periods of cleaning. Consult the food regulation in your jurisdiction to determine what lighting levels are required for your store.

Light bulbs must be shielded, coated or otherwise shatter-resistant when used in areas where there is exposed food; clean equipment, utensils, and linens; or unwrapped single-service and single-use utensils (Figure 5.23). This is to prevent glass fragments from getting into food and onto food-contact surfaces. Shielded or shatter-resistant bulbs are not required in areas where packaged foods are stored provided the packages will not be affected by broken glass falling onto them, and the packages are capable of being cleaned of debris from broken bulbs before the packages are opened.

Light bulbs over self-service buffets and salad bars must also be shielded or shatter-resistant. In addition, this lighting must not mislead customers about the quality of the food items on display.

a. Glass and food don't mix

b. Plastic coated shatterproof bulbs

c. Shielded lighting over exposed foods

Figure 5.23 Protect Food, Equipment, Linens, and Unwrapped Single-use Utensils from Broken Glass

[(a) and (b) used with permission of Shat-R-Shield, Inc.]

Heating, Ventilation, and Air Conditioning (HVAC)

An adequate supply of fresh air, suitable temperature, and humidity are necessary for employee comfort and productivity. An employee's comfort and well-being also require air that is free of obnoxious odors and other objectionable or injurious elements. This may pose a challenge in retail food establishments when heat, odors, and high air moisture are present.

Air conditioning in retail food establishments means more than simply "cooling the air." It includes heating, humidity control, circulation, filtering, and cooling of the air. HVAC systems must filter, warm, humidify, and circulate the air in the winter and maintain comfortable air temperature in the summer.

Ventilation in food production and warewashing areas is typically provided by mechanical exhaust hood systems. These systems keep rooms free of excessive heat, steam, condensation, vapors, obnoxious odors, smoke, and fumes. The standard ventilation system used in retail food establishments consists of a hood, fan, and intake and exhaust air ducts and vents (Figure 5.24). Ventilation hood systems must be designed and constructed to prevent grease or condensation from dripping onto food, equipment, utensils, linens, or single-service and **single-use articles**. The capacity of the ventilation system should be based on the quantity of vapor and hot air to be removed. The National Fire Protection Association recommends all hoods be equipped with a fire extinguishing system.

Figure 5.24 Ventilation Hood System

Hoods are usually constructed of stainless steel or a comparable material that provides a durable, smooth, and easily cleanable surface. The hood should be equipped with filters or other grease extracting equipment to prevent drippage onto food.

Filters and other grease-removal equipment must be:

- Easily removed for cleaning and replacement, or
- Designed to be cleaned in place.

Fires at retail food establishments have been caused by the combustion of grease that accumulates in filters and ducts. Cleaning and replacement of filters must be part of your routine maintenance program. Intake and exhaust air ducts must be cleaned so they do not become a source of contamination of dust, dirt, and other materials. If vented to the outside, ventilation systems must not create a public health nuisance or unlawful discharge.

Summary

In the case study presented at the beginning of this chapter, the glass from the bulb is a physical hazard that can contaminate food and food-contact surfaces. Customers could be seriously injured if they consumed food that contained pieces of glass. To protect your customers, the food that was exposed to the falling pieces of glass must be discarded, and contaminated food-contact surfaces must be thoroughly cleaned and sanitized.

To prevent similar situations in the future, maintenance personnel must be instructed not to replace burned out light bulbs while food-handling activities are being conducted on the counter below. If it was necessary to replace the bulb immediately, the deli employees should have removed the food and containers from the counter area. Otherwise, the maintenance person should wait until the deli employees were finished working with exposed and unpackaged foods.

Good planning is essential for maximum productivity and efficiency. The total allotted space needed and the arrangement of equipment in the space are the overall features you must consider when planning the layout of a retail food establishment. The design of your facility should promote the smooth flow of work, provide adequate facilities and space for performing work efficiently, and contribute to high sanitation standards.

The basic needs of a retail food establishment will determine the type and size of equipment purchased. When purchasing equipment, look for items that:

- ▲ Will serve the current and future needs of the operation
- ▲ Can be properly cleaned and sanitized
- ▲ Do not require extraordinary maintenance and repair.

The *FDA Food Code* and construction standards from the American Standards Institute (ANSI) require food equipment and utensils to be:

- ▲ Smooth
- ▲ Easily cleanable
- ▲ Easy to take apart
- ▲ Easy to reassemble
- ▲ Equipped with rounded corners and edges.

Materials used as food-contact surfaces must be safe, durable, corrosion-resistant, nonabsorbent, and resistant to chipping, pitting and deterioration. These materials must not allow the transfer of harmful substances, colors, odors, or tastes to food.

Always install equipment in accordance with local building, plumbing, electrical, health, and fire safety codes. The best possible design and construction of equipment is worthless if electrical, gas, water, or drain connections were inadequate or poorly installed.

Discussion Questions (Short Answer)

1. In what ways do layout, design, and facilities planning influence a retail food establishment?
2. What is a work center? What layout features are important to the efficiency of a work center?
3. What design and construction factors should a manager consider before buying a particular piece of equipment?
4. What standards organizations are commonly involved in the testing of food equipment? How can these organizations be helpful to a retail food establishment manager?
5. What kinds of materials are commonly used in the construction of equipment and utensils? What are the advantages and disadvantages of each kind of material?
6. How does a food-contact surface differ from a non-food-contact surface?
7. What are some features that make a piece of equipment easily cleanable?
8. How do proper lighting and ventilation affect sanitation in a retail food establishment?

Quiz 5 (Multiple Choice)

1. The design, layout and facilities in a retail food establishment should be based on the:
   a. Number of employees working in the store.
   b. Amount of food sold each day.
   c. Type of foods prepared and sold.
   d. Amount of space available in food production areas.
2. Which of the following organizations does not routinely evaluate the design and construction of food equipment?
   a. The U.S. Food and Drug Administration.
   b. NSF International.
   c. The American Gas Association.
   d. Underwriters Laboratories, Inc.
3. Which of the following criteria is least important when determining a retail food establishment's need for a particular piece of equipment? Will the equipment:
   a. Result in improved quality of food?
   b. Produce a significant savings in labor and material?
   c. Improve sanitation?
   d. Make the facility more attractive to customers?
4. When buying equipment, retail food establishment managers should look for equipment that:
   a. Can meet the need and demands of the operation.
   b. Can be used for a long time.
   c. Does not require excessive repair and upkeep.
   d. All of the above.

5. Which of the following statements is false?
    a. The size of equipment should be such that it can easily be fitted in the space available in the layout of the facility.
    b. Construction requirements for equipment vary according to whether the surface or area is a "food-contact surface" or a "splash zone."
    c. Wood is a preferred material for food-contact surfaces of equipment because it resists moisture and bacterial growth.
    d. Floor-mounted equipment that is not easily movable must be elevated on 6-inch legs or sealed to the floor on a base that is sealed to the floor.
6. Which of the following statements about ovens is false?
    a. A convection oven cooks food by circulating heated air around the product.
    b. Newer microwave ovens are sized to permit quantity cooking in a retail food establishment.
    c. The range oven is commonly used in retail food establishments.
    d. Deck ovens contain separate heating elements and controls for each unit.
7. Which of the following statements about refrigeration equipment is false?
    a. The size of refrigerator or freezer needed depends on whether a walk-in unit is available, what is going to be stored, and how much.
    b. Mechanical refrigeration during transportation and storage is important to preserve perishable foods.
    c. Pass-through refrigeration units open from both sides and are used for storage between production and service areas.
    d. The cleanliness and care of refrigerated equipment is not important, since it is only used to store and display foods.

8. What type of equipment is required when a retail food establishment uses a manual dishwashing operation?
    a. A three- or four-compartment sink with hot and cold potable running water and drainboards or easily movable dish tables.
    b. A two-compartment sink with hot and cold potable running water and drainboards or easily movable dish tables.
    c. A two-compartment sink with suitable facilities for pre-flushing and pre-scraping the dishes before they are washed.
    d. A single-tank dishwashing machine large enough to accommodate the largest pieces of equipment and utensils used in the operation.
9. From a sanitation perspective, what is the most important reason for having adequate lighting in a food production area?
    a. To provide a comfortable work environment for employees.
    b. To show when a surface is soiled and when it has been properly cleaned.
    c. To decrease accidents and waste due to employee error.
    d. To reduce glare that causes eye fatigue.
10. Which of the following statements about ventilation is false?
    a. Ventilation is typically provided by means of a mechanical exhaust hood system.
    b. Ventilation hood systems must be designed and constructed to prevent grease or condensation from dripping onto food and food-contact surfaces.
    c. The hood should be equipped with filters or other grease catching devices to prevent drippage into food.
    d. Intake and exhaust air ducts do not need cleaning if filters or other grease catching devices are provided to collect grease and other condensation.

Answers to the multiple choice questions are provided in Appendix A.

References/Suggested Readings

Food and Drug Administration and Conference for Food Protection (1997). *Food Establishment Plan Review Guide.* Washington, DC.

Food and Drug Administration (2001). *2001 Food Code - Chapter 4.* United States Public Health Service. Washington, DC.

Food and Nutrition Service (1999). *A Guide for Purchasing Food Service Equipment.* United States Department of Agriculture. Washington, DC.

Longreè, K., and G. Armbruster (1996). *Quantity Food Sanitation.* Wiley Interscience. New York, NY.

Suggested Web Sites

AK Steel Stainless Steel
www.aksteel-com/markets/stainless_steels.asp

NSF International
www.nsf.org

Underwriters Laboratories, Inc.
www.ul.com

Hobart Corporation
www.hobartcorp.com

Hussmann Company
www.hussmann.com

Katchall Industries International
www.KatchAll.com

Shat-R-Shield Inc.
www.shat-r-shield.com

Specialty Steel Industry of North America
www.ssina.com

The Food Marketing Institute
www.fmi.org

Gateway to Government Food Safety Information
www.foodsafety.gov

United States Department of Agriculture (USDA)
www.usda.gov

United States Food and Drug Administration (FDA)
www.fda.gov

USDA Foods Safety and Inspection Service (FSIS)
www.fsis.usda.gov

6 Cleaning and Sanitizing Operations

Children Suffer Allergic Reaction to Walnuts in Muffins

Two children were taken to the emergency room of a local hospital when they experienced allergic reactions after eating apple muffins prepared in the bakery at a local retail food establishment. The children's symptoms included red, itchy skin; swelling in their lips; and difficulty breathing.

The children's parents knew each child was allergic to walnuts. However, walnuts were not supposed to be an ingredient in the apple muffins the children ate. Samples of the suspect muffins were tested and traces of nuts were found in the product.

The emergency room physician contacted the local health department, and an investigation of the incident was conducted. Health department personnel conducted an inspection of the bakery and interviewed the employees who were on duty the day the contaminated muffins were made. The recipe for the apple muffins was reviewed, and it was verified that walnuts were not an ingredient in that product. However, interviews with bakery employees revealed the same mixing bowl had been used to make the batter for both banana nut muffins and apple muffins. The batter for the banana nut muffins was made first. Employees rinsed and dried the mixing bowl between batches.

How did the walnuts get into the apple muffins? What must the bakery employees do to prevent similar episodes from occurring in the future?

Learning Objectives

After reading this chapter you should be able to:

- ▲ Recognize the difference between cleaning and sanitizing.
- ▲ Identify the different processes that can be used to clean and sanitize equipment and utensils in a retail food establishment.
- ▲ Identify the primary steps involved in manually and mechanically cleaning and sanitizing equipment and utensils.
- ▲ Describe the factors that affect cleaning efficiency.
- ▲ Identify the procedures used to clean environmental areas in a retail food establishment.

Essential Terms

Cleaning agent

Clean-in-place (CIP)

Detergent

In-place sanitizers

Sanitize

Sanitizer

- **Chlorine**
- **Iodine**
- **Iodophors**
- **Quaternary Ammonium Compounds (Quats)**

Selectivity

Soap

Principles of Cleaning and Sanitizing

Cleaning and sanitizing are important activities in all retail food establishments. Proper cleaning and sanitizing enhance the safety and quality of food and increase the life expectancy of equipment and facilities.

Cleaning and sanitizing are two distinct processes used for very different purposes. Cleaning is the physical removal of soil from surfaces of equipment and utensils. Most of the soil is food wastes and residues. **Soil can also include waste materials such as plastic, paper, etc. Sanitizing** (sometimes called sanitization) **is the treatment of a clean surface to reduce the number of disease-causing microorganisms to**

safe levels (Figure 6.1). The equipment and supplies used for cleaning are different from those used for sanitizing.

Figure 6.1 Cleaning and Sanitizing

Effective cleaning consists of four separate events:

1. A detergent or other type of cleaner is brought into contact with the soil.
2. The soil is loosened from the surface being cleaned.
3. The loosened soil is dispersed in the wash water.
4. The dispersed soil is rinsed away along with the detergent to prevent it from being redeposited onto the clean surface.

Removal of Food Particles

Food residues should be scraped and flushed from the surfaces of equipment and utensils before the items are placed in a cleaning solution (Figure 6.2). Use warm water when pre-flushing and pre-scraping equipment and utensils. Avoid using very hot water or steam because they tend to "bake" food particles onto the surface of equipment and utensils and that makes cleaning more difficult.

Figure 6.2 Pre-Flushing Makes Washing Easier

Application of Cleaning Agents

A cleaning agent is a chemical compound formulated to remove soil and dirt. **Cleaning agents** typically include an acid or alkaline detergent and may include degreasers, abrasive materials or a sanitizer. It is important to choose the cleaning agent that is right for the particular job you have. Cleaners must be effective at removing soils from surfaces of equipment, utensils, floors, etc. They should also be safe and economical to use.

Soaking

Small equipment, equipment parts, and utensils may be immersed in cleaning solutions in a sink. By soaking equipment or utensils for a few minutes before scrubbing, you will increase the effectiveness of manual and mechanical warewashing.

Spray Methods

Cleaning solutions can be sprayed on equipment surfaces by using either fixed or portable spray units that use hot water or steam. These methods are used extensively in meat departments of retail food establishments (Figure 6.3).

Figure 6.3 Spray Cleaning

Clean-in-Place Systems

The clean-in-place method is an automated cleaning system generally used in conjunction with permanent-welded pipeline systems such as milk pasteurization equipment. The strength and velocity of the cleaning solution moving through the pipes are chiefly responsible for removing soil in clean-in-place (CIP) operations.

Abrasive Cleaning

Abrasive cleaners, in the form of powders and pastes, are used to remove soil that is firmly attached to a surface. Always rinse these cleaners completely and avoid scratching the surface of equipment and utensils. Abrasive type cleaners are not recommended for use on stainless steel surfaces. Never use metal or abrasive scouring pads on food-contact surfaces because small metal pieces from the pads may promote corrosion or may be picked up in food to become a physical hazard.

Rinsing

Immediately after cleaning, thoroughly rinse all equipment surfaces with hot, potable water to remove the cleaning solution. This very important rinse step is necessary because the product or detergent used for washing can interfere with the effectiveness of the sanitizer.

Factors Affecting Cleaning Efficiency

Many factors affect the efficiency of a cleaning process in terms of how well and how easily soil is removed. Among the most important factors are:

- ▲ Type of soil to be removed
- ▲ Water quality
- ▲ The detergent or cleaner to be used
- ▲ Water temperature
- ▲ Water velocity or force
- ▲ Time detergent remains in contact with the surface
- ▲ The concentration of cleaner.

Each of these factors is important for effective cleaning and can be modified to meet the needs of a particular type of soil or set of conditions.

Type of Soil to Be Removed

Soil may consist of:

- ▲ Food deposits (proteins and carbohydrates)
- ▲ Mineral deposits (salts)
- ▲ Microorganisms (bacteria, viruses, yeasts, and molds)
- ▲ Fats and oils
- ▲ Dirt and debris.

When you know the type and condition of the soil you are working with, it is easier to determine which cleaning agents and processes will remove that soil most effectively and economically.

Water Quality

Water is the primary component of cleaning materials used in retail food establishments. **The water supply serving an establishment must be safe to drink (potable). Potable water is free from harmful**

microorganisms, chemicals, and other substances that can cause disease. While the water supply must be potable, it may contain substances that bring about hardness, taste, and odors. Therefore, cleaning agents must be compatible with the characteristics of your water supply and the nature of your operation.

"Hard" water is caused by dissolved salts of calcium, magnesium, and iron. Water hardness reduces the effectiveness of detergents and leaves "lime" scale deposits on the surface of equipment. The degree of water hardness varies considerably from place to place. In severe cases, hard water must be softened to permit effective cleaning. If the water is not softened, cleaners should contain softening agents that inactivate the iron and manganese in the water without settling them out.

If there is an extended interruption in the water supply to a retail food establishment, the facility should cease operations or sell only prepackaged or preprepared food using single-use utensils until water service is restored.

Detergents and Cleaners to Be Used

A **detergent** is defined as a cleaning or purifying agent – a solvent that dissolves dirt and soil. The origin of the word is from the Latin, *detergeo*, meaning "to wipe away." Water acts as a detergent when soils are readily soluble. However, we can improve the cleansing action of water by adding soap, alkaline detergents, acid detergents, degreasers, abrasive cleaners, detergent sanitizers, or other cleaning agents to it.

Soaps

Bar **soaps** are cleaning agents made by the chemical reaction of alkali on fats or fatty acids. Soaps are only used as cleaners in retail food establishments in a limited way. In soft water, they are effective for washing hands. However, most soaps do not dissolve in cold water and are not compatible with some sanitizers. In hard water, soaps form troublesome precipitates and films that lose their cleaning power and break down into fatty acids. Bar soap has largely been replaced by synthetic detergents in retail food establishments.

Alkaline Detergents

An alkali is the principal detergent ingredient of most cleaners. Alkalis combine with fats to form soaps and with proteins to form soluble compounds easily removed by water. Some alkalis are good buffers and enhance detergency. However, in varying degrees, alkalis corrode aluminum, galvanized metal, and tin.

Sodium hydroxide (caustic soda or lye) is the strongest and cheapest of the alkalis. It has high detergency but is also highly corrosive for nearly all surfaces, including skin. Sodium carbonate is a mild alkali that has moderate dissolving power and is much less corrosive than strong alkalis. Some skin contact with sodium carbonate is possible.

Alkaline detergents are good general-purpose cleaners commonly used to clean equipment, utensils, floors, walls, and ceilings in retail food establishments.

Acid Detergents

Acid cleaners dissolve mineral deposits, such as calcium and magnesium precipitates, from equipment surfaces. Acid detergents are frequently used to remove food and hard water deposits from the surfaces of equipment and utensils.

The acids may be categorized into two groups: inorganic and organic. The inorganic acids are called mineral or strong acids. The strong acids are extremely corrosive to metals and should be used only in special cases. Cleaners often contain phosphoric acid to remove hard water film. Organic acids are not as corrosive to metals and are less irritating to the skin.

Degreasers

Degreasers are specialty products that remove grease and greasy or oily soil. Degreasers contain surfactants that penetrate and break up grease and oil. Degreasers are designed more for hard surfaces than for fabrics. They may be used in meat departments and other areas for pretreatment or as the sole cleaning agent. Equipment, utensils, and facilities should be thoroughly rinsed after degreasers have been applied.

Abrasives

Abrasives, when mixed with a detergent, are useful for jobs that require scrubbing, scouring, or polishing. Such naturally occurring mineral abrasives as pumice, quartz, and sand are ground into small particle sizes and supply scouring and polishing action to cleansers, hand soaps, and soap pads.

Abrasives should be used with care since they can cause scratches on metal surfaces, including stainless steel. In addition, the particles of the cleaner, along with metal particles scraped from equipment, may also contaminate food.

Detergent Sanitizers

As the name would suggest, detergent sanitizers are compounds that contain both a detergent and a sanitizer. These products effectively clean and sanitize a food-contact surface, provided a two-step process is used. That is, the detergent sanitizer must be applied to a food-contact surface two times. The first time is to clean the surface and a second time is to sanitize it. Food-contact surfaces should be thoroughly rinsed to remove chemical residues that may remain on the surface after the cleaning and sanitizing process.

A summary of the advantages and disadvantages of the cleaning agents most commonly used in retails food establishments is presented in Figure 6.4.

CLEANING AGENT	ADVANTAGES	DISADVANTAGES
Soaps	Effective for handwashing	May form precipitates and films in hard water and are not compatible with some sanitizers
Alkaline Detergents	Good general-purpose cleaners. Dissolve proteins and other organic material	Highly corrosive to equipment surfaces
Acid Detergents	Effective removal of mineral deposits	Strong inorganic acids are corrosive to metals and irritating to skin
Degreasers	Remove grease and oily soils from hard surfaces	Equipment must be thoroughly rinsed after degreasers have been applied
Abrasives	Useful for jobs that require scrubbing, scouring, or polishing	Can scratch equipment surfaces

Figure 6.4 Advantages and Disadvantages of Different Types of Detergents and Cleaners

Water Temperature

Heat-stable detergents work best when the water temperature of the solution is between 130°F (54°C) and 160°F (71°C). The temperature of the wash water needs to be hot enough to effectively remove soil, but it should not be so hot the soil will be "baked" onto the food-contact surface. Increased wash water temperatures help decrease the strength of the bonds that hold soil to the surface being cleaned.

Some detergents are designed specifically for cold water. Check the manufacturer's directions to determine the desired water temperature. A good rule of thumb is to use cold-water detergents when the wash water is at "tap-water" temperature.

Velocity or Force

In manual cleaning procedures, force is applied by "elbow grease." For mechanical and clean-in-place systems, the velocity and force of the cleaning solution help remove soil and film from food-contact surfaces. The importance of velocity and force on cleaning decreases as the effectiveness of the detergent increases. In other words, less "elbow grease" will be required when the detergent is formulated to effectively remove soil from the surfaces of equipment and utensils.

Amount of Time the Detergent or Cleaner Remains in Contact with the Surface

Cleaning efficiency is increased by use of longer contact times. The amount of scrubbing necessary to remove soil can be decreased by "soaking" items in detergents or cleaners before cleaning.

Concentration of the Detergent or Cleaner

Increasing the strength of a detergent increases the reaction rate and magnifies its cleaning power. However, there is an upper limit beyond which higher concentrations neither increase nor decrease efficiency. Always follow the manufacturer's recommendations for detergent concentration. Using too much detergent does not clean better or faster, and it is a waste of money.

Cleaning Frequency

Under normal circumstances, food-contact surfaces and equipment used to prepare and serve potentially hazardous foods must be cleaned throughout the day to prevent the growth of microorganisms on those surfaces. The *FDA Food Code* requires food-contact surfaces of equipment and utensils used with potentially hazardous foods to be cleaned at least every four hours at room temperature. Some additional guidelines for cleaning food-contact surfaces include:

- ▲ Before each use with a different type of raw animal food such as beef, fish, lamb, pork, or poultry

  [This requirement does not apply to situations where a food-contact surface or utensil is in contact with a series of different raw animal foods each requiring a higher cooking temperature. For example, equipment or utensils would not have to be cleaned when preparing raw fish followed by cutting raw poultry on the same cutting board. The final cooking temperature for poultry is 165°F (74°C) and is higher than the 145°F (63°C) final cooking temperature required for fish.]

- ▲ Each time there is a change from working with raw foods to working with ready-to-eat foods
- ▲ Between uses with raw fruits and vegetables and with potentially hazardous food
- ▲ Before using or storing a food temperature-measuring device
- ▲ At any time during the operation when contamination may have occurred.

Room Temperature	Cleaning Frequency
41°F (5°C) or less	at least once every 24 hours
41°F (0°C)- 45°F (7°C)	at least once every 20 hours
45°F (7°C)- 50°F (10°C)	at least once every 16 hours
50°F (10°C) - 55°F (12°C)	at least once every 10 hours

Figure 6.5 Room Temperature and Cleaning Frequency
(Source: *FDA Food Code*)

The *FDA Food Code* allows some exceptions to the four-hour rule for cleaning frequency. Surfaces of utensils and equipment that come into contact with potentially hazardous food may be cleaned less frequently than every four hours if:

- ▲ During storage, containers of potentially hazardous food and their contents are maintained at proper hot- and cold-holding temperatures and the containers are cleaned when they are empty.
- ▲ When utensils and equipment are used to prepare potentially hazardous food in a refrigerated room or area. The room

temperature and cleaning frequency prescribed in the *FDA Food Code* are found in Figure 6.5. The retail food establishment must maintain records to demonstrate proper cleaning frequency based on the ambient temperature of the refrigerated room or area. In all instances, equipment and utensils must be cleaned at least once every 24 hours.

- ▲ Containers used in serving situations such as salad bars, delis, and cafeteria lines are maintained at proper hot- and cold-holding temperatures and the content of the container is periodically combined with additional supplies of the same food that is at the required temperature, and the containers are cleaned at least once every 24 hours.
- ▲ Temperature-measuring devices are maintained in contact with food, such as when left in a container of deli food or in a roast, held at proper temperatures.
- ▲ Equipment is used for storage of packaged or unpackaged food such as a reach-in refrigerator and the equipment is cleaned at a frequency necessary to preclude accumulation of soil residues.
- ▲ In-use utensils are intermittently stored in a container of water in which the water is maintained at 140°F (60°C) or more and the utensils and container are cleaned at least every 24 hours or at a frequency necessary to preclude accumulation of soil residues.

Iced tea dispensers, carbonated beverage dispenser nozzles, water dispensing units, ice makers and ice bins are examples of equipment that routinely comes into contact with food that is not potentially hazardous. These types of equipment must be cleaned on a routine basis to prevent the development of slime, mold, or soil residues that may contribute to an accumulation of microorganisms. The *FDA Food Code* recommends surfaces of utensils and equipment contacting food that is not potentially hazardous be cleaned:

- ▲ At any time when contamination may have occurred.
- ▲ At least every 24 hours for iced tea dispensers and customer self-service utensils such as tongs, scoops, or ladles.
- ▲ Before restocking customer self-service equipment and utensils such as condiment dispensers and bulk food display containers.
- ▲ Whenever possible, follow the manufacturer's guidelines for regular cleaning and sanitizing of the food-contact surfaces of equipment and utensils. If the manufacturer does not provide cleaning instruction, the person in charge should develop a cleaning

regimen that will effectively remove soil, mold and other contaminants from equipment and utensils.

Sanitizing Principles

Heat and chemicals are the two types of sanitizers most commonly used in retail food establishments. Sanitizers **destroy disease-causing microorganisms that may be present on equipment and utensils even after cleaning.** Sanitizing does not provide the same level of microbial destruction as sterilization, because some bacterial spores and a few highly resistant vegetative cells generally survive the sanitizing process.

In all instances, a food-contact surface must be thoroughly cleaned and rinsed to remove soil and detergent residues before it can be properly sanitized.

Heat Sanitizing

Heat has several advantages over chemical sanitizing agents because it:

- Can penetrate small cracks and crevices
- Is noncorrosive to metal surfaces
- Kills all types of microorganisms equally effectively
- Leaves no residue
- Is easily measurable.

Heat destroys vegetative bacteria cells by disrupting some of the protein molecules in the cells. Moist heat is much more efficient in killing bacteria and other disease-causing microorganisms than dry heat. If you stick your hand in boiling water, you will burn yourself. However, if you stick your hand in an oven set at 450°F, it doesn't seem to be that hot. The reason the boiling water seems so much hotter is water is much more effective than air in transferring heat. Heat sanitizing is used in both manual and mechanical warewashing operations.

Heat sanitizing in manual warewashing operations requires cleaned equipment and utensils to be submersed in hot water maintained at 171°F (77°C) or above for at least 30 seconds. Manual warewashing operations rarely employ this type of sanitizing due to concerns about employee safety and the large amount of energy required to keep the water hot. If a retail food establishment uses hot water for sanitizing, it must provide a properly calibrated thermometer to routinely check the temperature of the sanitizing water. Employees must use dish baskets or racks to lower equipment and utensils into the sanitizing water. This allows the items to be completely covered with the hot water without the employee being scalded. Employees must use extreme care when working with water at these very high temperatures. Managers must enforce strict safety procedures and provide personal protective equipment for employees who are involved in manual warewashing operations.

Sanitizing with hot water is most commonly performed as part of mechanical warewashing operations. The *FDA Food Code* requires the hot water used for these purposes to be between 180°F (82°C) and 194°F (90°C) when it leaves the final rinse spray nozzles. This assures the temperature of the water will be at 171°F (77°C) or above when it reaches the surfaces of the equipment and utensils being sanitized. The only exception to these temperature requirements is in the single-tank, stationary-rack, and single-temperature machines, where the final rinse water temperature must be at least 165°F (74°C) as it leaves the spray nozzles.

Steam is an excellent agent for treating pieces of food equipment. The equipment shown in Figure 6.6 is an example of a low pressure, high temperature steam/vapor cleaning system. One of the most attractive features of this machine is the low amount of water it uses while cleaning. With the aid of this equipment, it is possible to clean food-contact surfaces of equipment without the use of chemicals. When steam in a flow cabinet is used, it should be sufficient to achieve 171°F (77°C) for at least 15 minutes or 200°F (94°C) for at least five minutes.

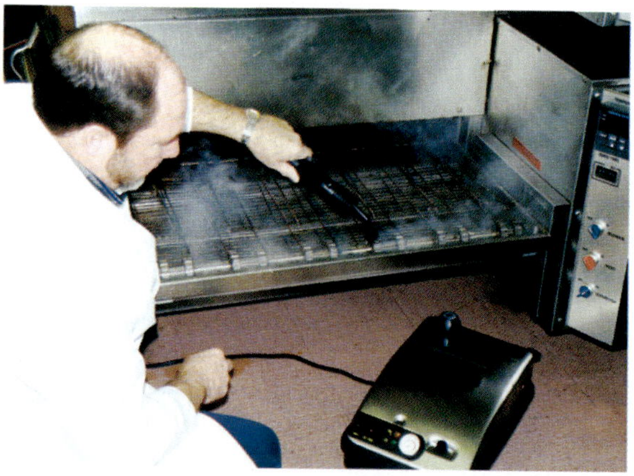

Figure 6.6 Low Pressure, High Steam Cleaning System

Reprint with Permission of Food Safety Academy, Skokie, IL and Hi-Tech Cleaning Systems

Steam is sometimes produced in remote steam boilers. Most boiler systems will require the periodic addition of boiler water additives to prevent the buildup of scale and corrosion in the boiler system. When there is a chance that steam produced in these boilers will come into direct or indirect contact with food, the chemicals added to boiler water must be approved for use as a food additive and labeled for that use as per Title 21 CFR Part 173.310 - Boiler Water Additives.

When using heat for sanitizing, the temperature at the surface of the equipment and utensils must reach at least 160°F (71°C) to ensure proper destruction of disease-causing microorganisms. The temperature at the surface can be measured by using irreversible heat-sensitive labels or tapes that are attached to a clean dry dish and then run with a load of soiled dishes through a regular wash cycle. The labels change from white to black when the indicated temperature is reached. Some examples of heat sensitive labels are illustrated in Figure 6.7. Temperature-sensing dishwashing labels are also available from All QA Products Company and Paper Thermometer Company.

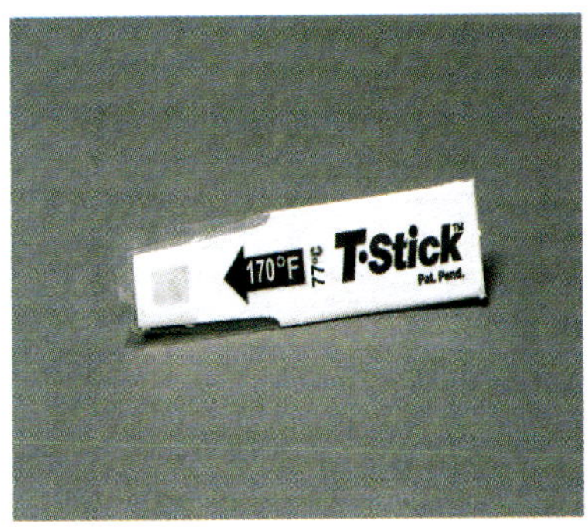

Figure 6.7 Color-Changing Labels for Measuring the Surface Temperature of Dishware

Another commonly used device for measuring the temperature of hot water sanitizers is the "maximum registering" or "holding" thermometer. This type thermometer will continue to hold the highest temperature measured until shaken down like a medical thermometer. An example of a maximum registering thermometer is presented in Figure 6.8. These instruments are becoming less popular because they contain mercury which can be a contaminant of food and the general enviroment.

Figure 6.8 Maximum Registering Thermometer
(Courtesy of Cooper Instrument Corporation)

Chemical Sanitizing

The chemical sanitizers most commonly used in retail food establishments are chlorine, iodine, and quaternary ammonium compounds. A summary of required sanitizer concentrations and proper conditions for the use of these chemicals is presented in Figure 6.9.

There are two ways to sanitize surfaces using chemical compounds. The first involves immersing a piece of equipment or a utensil into a sanitizing solution at the prescribed concentration. The second involves swabbing, brushing, or pressure spraying the sanitizing solution directly onto the surface of the equipment and utensils being sanitized.

The effectiveness of chemical sanitizers weakens over time as water from the wash and rinse stages of the process dilute the sanitizer. A chemical test kit or test strips must be available so warewashing personnel can routinely check the strength of the sanitizing solution. The sanitizing solution should be replaced with a fresh solution whenever it becomes contaminated or if the concentration falls below the minimum level recommended by the manufacturer of the sanitizer.

<table>
<tr><th>Minimum Concentrations of Sanitizer as expressed in parts per million (ppm) or mg/L</th><th>pH 10.0 or less & Minimum Water Temperature</th><th>pH 8.0 or less & Minimum Water Temperature</th><th>Contact Time</th></tr>
<tr><td>Chlorine 25</td><td>120°F (49°C)</td><td>120°F (49°C)</td><td>greater or equal to 10 seconds</td></tr>
<tr><td>Chlorine 50</td><td>100°F (38°C)</td><td>75°F (24°C)</td><td>greater or equal to 7 seconds</td></tr>
<tr><td>Chlorine 100</td><td>55°F (13°C)</td><td>55°F (13°C)</td><td>greater or equal to 10 seconds</td></tr>
<tr><td>Iodine greater or equal to 12.5 to 25</td><td colspan="2">pH greater or equal to 5.0 or per label; and water temperature at 75°F (24°C) or above</td><td rowspan="2">greater or equal to 30 seconds</td></tr>
<tr><td>Quaternary Ammonium, per label</td><td colspan="2">slightly alkaline pH is preferred, water hardness must be 500 ppm or less or per manufacturer's label, and water temperature at 75°F (24°C) or above</td></tr>
<tr><td colspan="4">NOTE: All chemical sanitizers shall be listed in 21 CFR178.1010 Sanitizing Solutions and used in accordance with the EPA-approved manufacturer's label use instructions.</td></tr>
</table>

Figure 6.9 Prescribed Conditions for Sanitizing Equipment and Utensils

Factors that Affect the Action of Chemical Sanitizers

The effectiveness of chemical sanitizers is affected by many different factors. Following are some of the most important factors to consider.

- **Contact of sanitizer** – in order for the chemical to react with and destroy microorganisms, it must achieve close contact.
- **Selectivity of sanitizer** – some sanitizers are nonselective and destroy a wide variety of microorganisms. Others exhibit a certain degree of selectivity. Chlorine, for example, is relatively nonselective. However, both iodophors and quaternary compounds have a selectivity that may limit their use in some retail food establishments.

- **Concentration of sanitizer** – in general, increasing the concentration of a chemical sanitizer proportionately increases its rate of microbial destruction. But there are limitations. The increased activity only extends to a certain maximum concentration, and then levels off so any further increase in concentration has no advantage. In rare instances, the destructive action may decrease with increased concentration. Concentration limits have been set for each type of sanitizer. More is not always better, and high concentrations of sanitizers can be toxic and wasteful. Always follow the manufacturer's label use instructions to ensure peak effectiveness of chemical sanitizers.
- **Temperature of solution** – all the common sanitizers increase in activity as the solution temperature increases. This is due in part to the fact chemical reactions are speeded up by raising the temperature. The recommended range of water temperatures for chemical sanitizing solutions is between 75°F (24°C) and 120°F (49°C). Under certain conditions, chlorine sanitizers may be used in conjunction with water that has a temperature as low as 55°F (13°C). This will be discussed later in this chapter. Quaternary ammonium sanitizers remain active in water that has a temperature above 120°F (49°C). However, chlorine and iodophor sanitizers lose their potency quickly when used in hot water.
- **pH or acidity of solution** – water hardness can affect the pH of water which exerts a significant influence on most sanitizers. Quaternary compounds react differently to changing pH, depending on the type of microorganism being destroyed. Chlorine and iodophors generally decrease in effectiveness with an increase in pH. Most soaps and detergents are alkaline with a pH between 10 and 12. That is why the soap or detergent must be rinsed off before a surface is sanitized.
- **Time of exposure** – allow sufficient time for chemical reactions to destroy the microorganism. The amount of exposure time depends on the preceding factors as well as the size of the microbial populations and their susceptibility to the sanitizer.

There are a wide variety of chemicals available to destroy or inhibit the growth of microorganisms. However, many of these chemicals are not suitable for use on food-contact surfaces because they may corrode, stain, or leave a film on the surface. Others may be toxic or too expensive for practical use. Therefore, the chemical sanitizing agents discussed in this section will be limited to those agents commonly used in retail food establishments. If you are considering using a chemical sanitizer other than chlorine, iodine, or a quaternary ammonium compound, you must make sure they have been approved and are safe to use on food-contact surfaces, and you must use them in accordance with the directions provided on the manufacturer's label.

Chlorine

Chlorine is a chemical component of hypochlorites. These compounds are commonly used as chemical sanitizers in retail food establishments. Hypochlorites offer many advantages as sanitizers. The chief advantages they offer are they:

- Control a wide range of microorganisms through germicidal action
- Deodorize and sanitize
- Are nontoxic to humans when used at recommended concentrations
- Are colorless and nonstaining
- Are easy to handle
- Are economical to use.

Hypochlorites are available as powders or liquids. Calcium hypochlorite is generally available in powder form and contains 70% available chlorine. Sodium hypochlorite, more commonly called household bleach, comes in a liquid form and contains between 2 and 6% available chlorine. Industrial strength hypochlorites usually contain between 10 and 18% available chlorine. Hypochlorites release hypochlorous acid in solution. It is the hypochlorous acid that provides chlorine's ability to destroy disease-causing microorganisms.

The germicidal effectiveness of chlorine-based sanitizers depends, in part, on water temperature and pH of the solution. The *FDA Food Code* requires food-contact surfaces to be exposed to a 50-ppm chlorine sanitizer for at least seven seconds when the chlorine solution has a pH of 10 or less and a water temperature of 100°F (38°C) or a pH of 8 or less and a temperature of at least 75°F (24°C). Equipment and utensils must be exposed to the sanitizer for at least 10 seconds if the strength, pH, and temperature requirements described above are not met. It is important to leave equipment and utensils in contact with the sanitizer long enough to significantly reduce the number of disease-causing microorganisms that remain on the surface after cleaning. State and local food codes may require equipment and utensils to be immersed in chlorine sanitizers for longer periods of time. Consult the food regulatory agency in your area to determine the sanitizing requirements for your establishment.

The effectiveness of hypochlorites is reduced by even small amounts of food soils and other types of organic matter. Therefore, a fresh water rinse must be provided to remove detergent and soils from equipment and utensils before immersing them in the sanitizing solution. A chlorine test kit or strips must be provided to measure the concentration of the chlorine sanitizer at least once per hour. The sanitizer solution must be replaced whenever it becomes contaminated or the concentration falls below the required strength.

Iodine

Iodine is chemically related to chlorine and has long been used to kill disease-causing microorganisms. The iodine-containing sanitizers commonly used in retail food establishments are called **iodophors** Iodophors are effective against a wide range of bacteria, small viruses and fungi. They are especially effective for killing disease-causing microorganisms found on human hands. Iodophors kill more quickly than either chlorine or the quaternary ammonium compounds.

Iodophors function best in water that is acidic and at temperatures between 75°F (24°C) and 120°F (49°C). Iodophors must be applied at 12.5 parts per million when immersion sanitizing and at 25 parts per million in swab and spray applications. Food-contact surfaces must be exposed to an iodophor for at least 30 seconds to ensure proper disinfection.

Iodophors are less influenced by organic matter than are hypochlorites. However, they are more expensive to use, and they will discolor and stain some surfaces such as silver, silverplate, and copper. Iodophors are also slippery and harder to handle than hypochlorites.

Quaternary Ammonium Compounds (quats)

Quaternary ammonium compounds (quats) are ammonia salts used as chemical sanitizers in retail food establishments. Quats are effective sanitizers, but they do not destroy the wide variety of disease-causing microorganisms that chlorine and iodophor sanitizers do. The *FDA Food Code* recommends using quats at 200 parts per million (ppm) for immersion sanitizing of food-contact surfaces. Quats are noncorrosive and have no taste or odor when used in the proper dilution. Quats are more heat stable than either hypochlorites and iodophors. Therefore, they work well at all temperatures above 75°F (24°C). They are effective in a wide pH range (although they are most effective in slightly alkaline water). The recommended contact time for quats is at least 30 seconds.

Quats should be used only in water with 500 ppm hardness or less, or in water having a hardness no greater than specified by the manufacturer's label.

At concentrations above 200 ppm, quats can leave a residue on the surface of an item. This is undesirable for food-contact surfaces. It is generally recommended quats not be used at concentrations above 200 ppm. Always follow the manufacturer's directions for proper uses and concentration of quats. You may also consult with the retail food regulatory agency in your jurisdiction about the uses of these and other chemical sanitizers.

A summary of the advantages and disadvantages of the chemical sanitizers used most frequently in retail food establishments is presented in Figure 6.10.

SANITIZER	ADVANTAGES	DISADVANTAGES
Chlorine compounds	Economical cost Kills many types of microbes Good for most sanitizing applications	Corrosive to equipment Can irritate human skin and hands
Iodophors	Less corrosive to equipment Less irritating to skin Good for killing microbes on hands	Moderate cost Can stain equipment
Quats	Stable at high temperature Stable for a longer contact time Good for in-place sanitizers	Very expensive Hard water can reduce effectiveness

Figure 6.10 Advantages and Disadvantages of Selected Chemical Sanitizers

Mechanical Warewashing

Mechanical warewashing is performed in a dishwashing machine that cleans and sanitizes multi-use equipment and utensils automatically (Figure 6.11). Dishwashing machines are designed to clean and sanitize equipment and utensils that have no electrical parts and will fit into the machine. When properly maintained and operated, mechanical warewashing is as effective at removing soil and disease-causing microoganisms as is manual warewashing.

Figure 6.11 Single-Tank Dishwashing Machine

Mechanical Warewashing Process

Mechanical warewashing uses an eight-step process to clean and sanitize equipment and utensils. A summary of the mechanical process is presented below:

1. Pre-scrape and pre-flush soiled equipment and pre-soak utensils to remove visible soil.
2. Rack equipment and utensils so wash and rinse waters will spray evenly on all surfaces and the equipment will freely drain.
3. Wash equipment and utensils in a detergent solution.
4. Rinse equipment and utensils in clean water at a temperature consistent with the type of dishwashing machine being used.
5. Rinse equipment and utensils in a fresh hot water sanitizing rinse between 180°F (82°C) and 194°F (90°C), except for a single-tank, stationary-rack, or single-temperature machine, where the final rinse may not be less than 165°F (74°C). The recommended final rinse temperature for low-temperature chemical sanitizing dishwashing machines is 120°F (49°C) or less.
6. Drain properly and air-dry equipment and utensils.

7. Store clean and sanitary items in a clean, dry area where they are protected from contamination.
8. Clean and maintain the machine to keep it in proper working condition.

The commercial dishwashing machines on the market today will work efficiently and effectively provided they are properly maintained and operated. Managers must accept responsibility for overseeing the operation and maintenance of this equipment when it is available in their establishment. When purchasing a dishwashing machine, always select a unit compatible with the types of equipment and utensils to be washed and the volume of equipment and utensils that must be processed during peak periods.

Provide sufficient counter or table space for the accumulation of soiled equipment and utensils and keep them separate from items that are clean and sanitary. Scrape food debris into a garbage disposal or garbage can. Freshly deposited soil is the material that remains on equipment and utensils after meal preparation or a meal itself. This is the easiest type of soil to remove because washing usually occurs while the soil is still moist.

Dried deposits of soil result when dirty dishes and utensils stand for a long period of time or are exposed to high temperatures. This allows fresh soft soil to dry, harden, and form a crusty deposit that is hard to remove. Dried soil frequently requires pre-soaking to soften and loosen it from the surface.

Once items are pre-flushed and pre-scraped, position them in a rack or on a conveyor so wash and rinse water and the sanitizer will spray evenly on all surfaces. Utensils and small equipment are racked with their bottom side up. Dishes are racked on their edge and eating utensils are placed in baskets with the mouthpiece up. After the utensils are washed, rinsed, and sanitized, invert them into another basket with the handles up and repeat the warewashing process.

The temperature of the wash water (and power rinse if applicable) is extremely important in providing effective sanitizing of equipment and utensils. Minimum wash and rinse water temperatures are provided in

Figure 6.12. Equipment and utensils should be exposed to the wash water for at least 40 seconds.

Type of Machine	Wash Temperature	Rinse Temperature
Single-tank, stationary-rack, single-temperature machine	165°F (74°C)	165°F (74°C)
Single-tank, conveyor, dual temperature machine	160°F (71°C)	180°F (82°C)
Single-tank, stationary-rack, dual temperature machine	150°F (66°C)	180°F (82°C)
Multi-tank, conveyor, multi-temperature machine	150°F (66°C)	160°F (71°C) to 180°F (82°C)
Chemical sanitizing machine	120°F (49°C)	120°F (49°C)

Figure 6.12 Minimum Wash and Rinse Temperatures of Mechanical Dishwashing Machines

(Source: *FDA Food Code*)

Some machines provide a power rinse that removes soil and detergent residues. The final rinse is designed to sanitize at a temperature of not less than 165°F (74°C) for at least 30 seconds in single-tank, stationary-rack, and single-temperature machines. Sanitizing in all other machines requires a final rinse temperature between 180°F (82°C) and 194°F (90°C) for at least 30 seconds. The flow pressure of the hot water sanitizing rinse must be between 15 and 25 pounds per square inch (psi). The reduced water pressure is required in order to assure the sanitizing water covers the entire surface of the items being sanitized.

Wash and power rinse water temperatures less than those prescribed in the *FDA Food Code* will not effectively sanitize equipment and utensils even when the final rinse (sanitizing) temperature is properly maintained. This is because effective sanitizing is the result of the cumulative temperature effects of wash, power rinse, and final rinse waters. Operators of dishwashing machines must be careful not to contaminate cleaned and sanitized equipment and utensils by touching them with soiled hands.

Cleaned and sanitized equipment and utensils must be properly drained and air dried before storage (Figure 6.13). Soiled, multi-use cloth towels can recontaminate cleaned and sanitized equipment and utensils. The *FDA Food Code* prohibits cloth-drying of equipment and utensils, with the exception that air-dried utensils can be "polished" with cloths that are clean and dry.

Figure 6.13 Air-Dry Utensils

Cleaned and sanitized equipment and utensils must be stored in such a way as to protect them from contamination (Figure 6.14). Cabinets and carts used to store cleaned and sanitized equipment and utensils may not be located in locker rooms and toilet rooms and under sewer lines that are not shielded to catch drips, leaking water lines, open stairwells, or under other sources of contamination.

Figure 6.14 Proper Storage of Cleaned and Sanitized Equipment and Utensils

Manual Warewashing

Manual warewashing typically uses a three-compartment sink to clean and sanitize equipment and utensils (Figure 6.15). The manual process begins with scraping and flushing food residues off the surface of equipment and utensils. When necessary, items may be pre-soaked to remove food residues and other soil. Pre-flushing and pre-scraping help keep the wash water in the first compartment of the sink clean.

After pre-flushing, the equipment and utensils are washed in the first compartment of the sink with warm water and a cleaning agent. Washing removes visible food particles and grease. Warewashing personnel should always use the correct amount of detergent based on the quantity of water used in the wash compartment of the sink. The temperature of the wash water must be maintained at not less than 110°F (43°C) unless a different temperature is specified by the manufacturer of the cleaning agent. A cloth, nylon brush, or other approved cleaning device can be used to loosen and remove soil. Soil and other deposits can shield bacteria from chemical sanitizing solutions. Therefore, always make certain surfaces are properly cleaned before they are sanitized.

Figure 6.15 Manual Warewashing

The second compartment of the sink is used to rinse equipment and utensils in clean, warm water. Rinsing removes cleaning agents, soap film, remaining food particles, and abrasives that may interfere with the sanitizer agent. For best results, the rinse water must be kept clean and the temperature should be between 110°F (43°C) and 120°F (49°C).

The third compartment of the sink is used for sanitizing. Items are sanitized with either hot water or chemical sanitizers. When hot water is used to sanitize, the temperature of the water must be maintained at 171°F (77°C) or above at all times. Items must be immersed in the hot water for at least 30 seconds. Chemical sanitizers are generally preferred in retail food establishments because they do not require large amounts of hot water, which saves energy. Refer to Figure 6.8 for *FDA Food Code* recommendations for the proper use of chemical sanitizers. Under certain conditions, chlorine sanitizers may be used with a water temperature as low as 55°F (13°C). Except for quats, water temperatures above 120°F (49°C) must be avoided because at high temperatures the potency of the chemical sanitizer is altered. State and local food codes may require equipment and utensils to be immersed in chemical sanitizers for at least 60 seconds. You should consult the regulatory authority in your jurisdiction to find out what rules apply in your area.

Each compartment of the sink must be large enough to permit the immersion of the largest pieces of equipment and utensils used in your operation. When equipment and utensils are too large to fit in the warewashing sink, they must be cleaned and sanitized in a mechanical dishwasher or by other manual operations such as high-pressure detergent sprayers, low- or line-pressure spray detergent foamers, or other task-specific cleaning equipment.

The manual warewashing sink must be equipped with sloped drain boards or dish tables of adequate size to store soiled items prior to washing, and clean items after they have been sanitized. Hot and cold potable water must be supplied to each compartment of the sink, and the sink should be cleaned and sanitized prior to use.

As with mechanical warewashing, items that have been manually cleaned and sanitized must be air-dried before they are put into storage. Avoid wiping cleaned and sanitized equipment and utensils with cloths or towels because this can recontaminate the sanitized surfaces. If drying agents are used in conjunction with sanitization they must contain components that are Generally Recognized as Safe (GRAS) for use in food, GRAS for the intended purpose, or have been approved for use as a drying agent under the relevant provisions contained in Title 21 of the Code of Federal Regulations.

Cleaned and sanitized equipment and utensils must be stored in such a way as to protect them from contamination. Shelves, cabinets and carts used to store cleaned and sanitized equipment and utensils may not be located in locker rooms and toilet rooms and under sewer lines that are not shielded to catch drips, leaking water lines, open stairwells, or under other sources of contamination.

The effectiveness of the manual warewashing process depends on the condition of the wash, rinse, and sanitizing solutions. In many retail food establishments, the detergent and sanitizer are dispensed into the sink automatically. In spite of this "automatic" feature, warewashing personnel should monitor the wash solution and replace it when it becomes soiled or the detergent is gone. When detergent builds up in the second compartment, the fresh water rinse should be drained and replaced with clean, fresh water. When the sanitizer in the third compartment is depleted, it should be drained completely and replaced with a fresh solution at proper strength. The concentration of a chemical sanitizer solution must be tested periodically with a test kit or test strips to make sure it remains at strength. These strips provide a color comparison to indicate the strength of the sanitizer (see Figure 6.16). Test kits or strips may be obtained from the manufacturer of your sanitizer.

Figure 6.16 Measuring the Concentration of Sanitizers Using Test Strips

Cleaning Fixed Equipment

Some of the equipment used in retail food establishments have food-contact surfaces but cannot be cleaned using either traditional mechanical or manual warewashing processes. This is called fixed equipment and common examples include band saws, floor-mounted mixers, slicers, grinders, produce display cases and the live tanks used to display live lobsters and mollusks (Figure 6.17). This equipment must be disassembled to expose food-contact surfaces to cleaning and sanitizing agents. The basic steps for cleaning fixed equipment are:

1. Disconnect the power supply to the equipment before taking it apart for cleaning. This will protect employees against electrocution and other types of injury.
2. Disassemble the equipment as necessary to allow the detergent solution to reach all food-contact surfaces.

3. Use a plastic scraper to clean equipment parts and remove food debris that has accumulated under and around the equipment. Scrape all debris into the trash.
4. Carry the parts that have been removed from the equipment to the manual warewashing sink where they will be washed, rinsed, and sanitized.
5. Wash the remaining parts of the equipment using a clean cloth, brush, or scouring pad and warm, soapy water. Clean from top to bottom.
6. Rinse thoroughly with fresh water and a clean cloth.
7. Swab or spray a chemical sanitizing solution, mixed to the manufacturer's recommendations, onto all food-contact surfaces.
8. Allow all parts to drain and air-dry.
9. Reassemble the equipment.
10. Resanitize any food-contact surface that might have been contaminated due to handling when the equipment was being reassembled.

Figure 6.17 Cleaning Fixed Equipment

Large equipment such as preparation tables and band saws can be cleaned by using a foam or spray method. In this process, detergents and degreasers, fresh water rinse, and a chemical sanitizer are applied using foam or spray guns. The hoses, feed lines, and nozzles that make up the foam or spray unit should be in good condition and attached properly.

The various display cases used in meat, seafood, dairy, bakery, produce and other departments in retail food establishments will have racks and other parts that can be removed for cleaning (Figure 6.18). It is important to remember food-contact surfaces have to be cleaned and sanitized while non-food-contact surfaces require cleaning only. Some equipment, such as produce and self-service bakery cases, may be equipped with accessory items such as misting heads and tongs or other handling devices. These accessories must also be cleaned and sanitized regularly to assure the safety and wholesomeness of products with which they are used.

Figure 6.18 Cleaning Large Equipment

Routine cleaning and maintenance of live seafood displays and holding tanks are necessary to preserve water quality and the quality of your live seafood. Some general guidelines that should be followed when working with this equipment are:

- ▲ Clean and sanitize the ultraviolet (UV) light unit used to kill pathogens in the water
- ▲ Clean the interior portion of the tank on a regular basis to keep it free of algae and slime buildup that may harbor pathogens
- ▲ If the tank is used for crab, lobster, etc., it must be cleaned and sanitized before oysters, clams, and other molluscan shellfish can be put into the tank.

Like any other equipment in your establishment, the live seafood display and holding tank should be checked on a regular basis to ensure it is in good working condition. Always follow the manufacturer's instructions when cleaning, sanitizing and maintaining your equipment.

The produce misting system in your establishment should be checked regularly to make sure it is clean and in good working order (see Figure 6.18). Routine maintenance should include the following steps:

- ▲ If there is a separate water filter for the misting system, it should be checked and replaced as needed to assure an adequate supply of clean water
- ▲ Disassemble the nozzles and clean and sanitize the various parts to remove food debris and mineral residue
- ▲ Flush the tubing that supplies the mister heads with clean water
- ▲ Wipe the outside surfaces of the system (tubing, etc.) with an approved sanitizer to stop the growth of fungi, molds, and yeast.

The bucket and brush method is often used to clean equipment that could be damaged by pressure spraying or immersion in the manual warewashing sink. This system uses one or more buckets for washing, rinsing, and sanitizing. Regardless of the number of buckets used, employees must make sure the equipment is properly cleaned and rinsed before it is sanitized.

The clean-in-place (CIP) method is used for equipment that is designed to be cleaned and sanitized by circulating cleaning and sanitizing solutions through the equipment. Examples of equipment that would be cleaned using this method are soft-serve ice cream or yogurt machines (Figure 6.19). The basic steps in the clean-in-place method are:

1. Empty food product and waste from the equipment.
2. Disconnect the power to the equipment.
3. Disassemble if parts are removable and cleanable in the manual warewashing sink.
4. Clean and sanitize the removable parts using the three-compartment warewashing sink and process.
5. Clean and sanitize the main unit by circulating wash, rinse, and sanitizing solutions through it. Combination detergent sanitizers can also be used on equipment that is designed for CIP cleaning.
6. Reassemble parts removed from the main unit.
7. Resanitize by circulating a manufacturer approved sanitizing solution through the equipment.

Figure 6.19 Soft-Serve Ice Cream Machine
(Courtesy of SaniServ)

Food-contact surfaces and equipment used to prepare potentially hazardous foods must be cleaned throughout the day to prevent the growth of disease-causing microorganisms. The *FDA Food Code* recommends this equipment is cleaned at least once every four hours or more frequently when used with different raw animal foods. Always follow the manufacturer's instructions for recommended cleaning frequency and procedures.

Many departments will use wiping cloths to clean up spills. These cloths should be stored in a container of sanitizing solution between uses in order to prevent microbial growth on the cloth (Figure 6.20). This is often referred to as an "in-place sanitizer." Any standard chemical sanitizer may be used for storing wiping cloths. However, quaternary ammonia compounds are preferred because they kill disease-causing microorganisms for a longer period of time. A clean cloth and fresh container of sanitizing solution at the proper strength must be used at the start of each day. The cloth should be changed whenever it becomes heavily soiled and the sanitizer must be changed when it falls below the required concentration. Wiping cloths may not be used for any other purpose except removing spills. Cloths used to wipe spills from floors and non-food-contact surfaces must be kept separate from the cloths used to clean food-contact surfaces of equipment.

Many retail food establishments use a variety of single-service items. These are typically foam or paper trays, individual packaging material, wrapping plastic or paper, and paper or plastic dishes, cups, and eating utensils that are used one time and then discarded. Single-use items must be stored and dispensed to protect them from contamination from employees, customers, and the surrounding environment. Dishes and cups should be stored in boxes or cabinets or upside down and away from overhead plumbing lines to protect them from contamination. Eating utensils should be either wrapped in plastic or stored in an approved dispensing container with the handles up. This will prevent the eating part of the utensil, the one that comes into contact with the customer's mouth, from becoming contaminated.

Figure 6.20 Store and Use Wiping Cloths Properly

Cleaning Environmental Areas

Food debris or dirt on floors, walls, ceilings, floor drains and other non-food-contact surfaces may provide a suitable environment for the growth of disease-causing microorganisms which employees may accidentally transfer to food. These surfaces must be cleaned regularly to minimize contamination, odors, and pest infestations (see Figure 6.21). A regular cleaning schedule should be established and followed to maintain the facility in a clean and sanitary condition. The manager or supervisor should create a master cleaning schedule that will list the following items:

- ▲ The specific equipment and facilities to be cleaned
- ▲ The processes and supplies needed to clean the equipment and facilities
- ▲ The prescribed time when the equipment and facilities should be cleaned
- ▲ The name of the employee who has been assigned to do the cleaning.

Figure 6.21 Cleaning Environmental Areas

Major cleaning should be done during periods when the least amount of food will be exposed to contamination and service will not be interrupted. The best time is when the establishment is not open for business. This will not be possible for operations that are open 24 hours per day and seven days a week. In these establishments, employees should minimize general purpose cleaning activities during periods when exposed or unpackaged foods are being prepared or handled. This requirement does not apply to cleaning that is necessary to remove spills or debris due to accidents.

Ceilings

Ceilings should be checked regularly to make certain they are not contaminating food production areas. Ceilings, lights, fans, and covers can be cleaned using either a wet or dry cleaning technique. When wet-cleaning ceilings and fixtures, it is best to use a bucket method to keep water away from lights, fans, and other electrical devices. Whenever possible, disconnect power before cleaning fixtures.

Walls

Some parts of the walls in a food production area, such as those around a sink or food production equipment, should be considered food-contact surfaces. These areas should be washed, rinsed, and sanitized whenever the other food-contact surfaces in the area are cleaned. Other wall areas are considered non-food-contact surfaces and do not need to be cleaned as often.

Walls may be cleaned using either the bucket or spray methods. However, walls should not be spray cleaned if spraying will damage the walls or contaminate exposed equipment and supplies in the area.

Floors

The floor is an important part of production, warewashing, and food storage areas. Floor construction and condition are important factors in maintaining a sanitary operation. Floors in all areas of a retail food establishment are subject to harsh treatment. Floor-cleaning machines, cleaning solutions, hot water, steam, scrubbers, food wastes, and heavy traffic all contribute to floor deterioration.

The cleanliness of floors depends on the ability of an employee to remove soil from the surface of the floor. Floors in customer shopping areas are usually cleaned with floor-cleaning machines, and floors in food production areas can be cleaned using a spray system for washing, rinsing, and sanitizing. Floors that will be damaged by spray cleaning can be cleaned using the bucket method. All floors should be sloped to drains to help remove soil and the water used to clean floors.

If floor drains are not properly maintained, they can be a source of disease, vermin, and odors. All floor drains should be flushed, washed, rinsed, and sanitized.

The steps commonly used to clean floor drains are listed below:

1. Remove the grate or cover over the drain.
2. Clean out waste and other debris from the drain.
3. Use a sprayer or hose to flush the drain and grate or cover.
4. Pour in drain cleaner to break up grease and other waste in the drain.
5. Wash the drain using a brush or water pressure from the sprayer.
6. Rinse the drain with hot water.
7. Pour or spray sanitizer into the drain.

Odor problems from floor drains can be solved by regularly flushing drains with hot water. This should be the last step in the daily cleaning program.

If not properly maintained, sewer pipes can become clogged and can cause sewage and other wastes to back up into and flood various areas of a retail food establishment. This can create both a health hazard and a nuisance. You must know what to do in the event a sewer backup occurs. Some of the steps you should follow are:

- Remove anything from the floor that might be damaged or contaminated by the waste water.
- Remove the blockage in the pipe that is causing the backup. This may require the services of a professional plumber.
- Wash and sanitize all equipment, floors, walls, and any other objects that may have been contaminated by the waste that backed up into the establishment.

Equipment and Supplies Used for Cleaning

Managers must make certain employees have the proper equipment and supplies needed to clean food-contact and nonfood-contact surfaces in retail food establishments. Some examples of equipment commonly used during cleaning are nylon brushes, cleaning cloths, scouring pads, squeegees, mops, buckets, spray bottles, hoses, and spray or foam guns. Commonly used cleaning supplies include hot water, cleaners, degreasers, and sanitizers.

There should be a separate sink to fill and empty mop buckets, to rinse and clean mops and to clean brushes and sponges. Handwashing, food preparation, and warewashing sinks must never be used for cleaning mops and brushes. A "janitor's" sink or floor drain should be provided to dispose of waste water produced by cleaning activities.

Containers of poisonous or toxic materials and personal care items must bear a legible manufacturer's label. Working containers used for storing poisonous or toxic materials such as cleaners and sanitizers taken from bulk supplies must be clearly and individually identified with the common name of the material. A container previously used to store poisonous or toxic materials may not be used to store, transport, or dispense food.

The Occupational Safety and Health Administration (OSHA) requires employees have the "Right-to-Know" about the hazardous chemicals to which they may be exposed on the job. Information about hazardous substances in the workplace is provided to employees by way of Material Safety Data Sheets (MSDS). Material Safety Data Sheets are developed by chemical manufacturers, and they must be maintained on file and accessible to employees in the food establishment (Figure 6.22).

Examples of information contained in an MSDS include:

- Ingredients
- Physical and chemical characteristics
- Fire, explosion, reactivity, and health hazard data
- How to handle hazardous chemicals safely
- How to use personal protective equipment and other devices to reduce risk
- Emergency procedures to use if required.

Failure to produce the required MSDS upon demand can result in a substantial fine from OSHA.

Material Safety Data Sheet

HEALTH	2*
FLAMMABILITY	0
REACTIVITY	1
Personal Protection	B

I Chemical Identification

NAME: REGULAR BLEACH

CAS no. N/A

DESCRIPTION: CLEAR, LIGHT YELLOW LIQUID WITH CHLORINE ODOR

RTECS no. N/A

Other Designations	Manufacturer	Emergency Procedure
EPA Reg. No. 5813-1 Sodium hypochlorite solution Liquid chlorine bleach Liquid Bleach	12 Broadway Broad Oak, CA. 94612	Notify your Supervisor Call your local poison control center or Rocky Mountain Poison Center (303) 573-1014

II Health Hazard Data

*Causes severe but temporary eye injury. May irritate skin. May cause nausea and vomiting if ingested. Exposure to vapor or mist may irritate nose, throat and lungs. The following medical conditions may be aggravated by exposure to high concentrations of vapor or mist: heart conditions or chronic respiratory problems such as asthma, chronic bronchitis or obstructive lung disease. Under normal consumer use conditions the likelihood of any adverse health effects are low. FIRST AID: EYE CONTACT: Immediately flush eyes with plenty of water. If irritation persists, see a doctor. SKIN CONTACT: Remove contaminated clothing. Wash area with water. INGESTION: Drink a glassful of water and call a physician. INHALATION: If breathing problems develop remove to fresh air.

III Hazardous Ingredients

Ingredients	Concentration	Worker Exposure Limit
Sodium hypochlorite CAS# 7681-52-9	5.25%	not established

None of the ingredients in this product are on the IARC, NTP or OSHA carcinogen list. Occasional clinical reports suggest a low potential for sensitization upon exaggerated exposure to sodium hypochlorite if skin damage (eg. irritation) occurs during exposure. Routine clinical tests conducted on intact skin with Liquid Bleach found no sensitization in the test subjects.

IV Special Protection Information

Hygienic Practices: Wear safety glasses. With repeated or prolonged use, wear gloves.

Engineering Controls: Use general ventilation to minimize exposure to vapor or mist.

Work Practices: Avoid eye and skin contact and inhalation of vapor or mist.

V Special Precautions

Keep out of reach of children. Do not get in eyes or on skin. Wash thoroughly with soap and water after handling. Do not mix with other household chemicals such as toilet bowl cleaners, rust removers, vinegar, acid or ammonia containing products. Store in a cool, dry place. Do not reuse empty container; rinse container and put in trash container.

VI Spill or Leak Procedures

Small quantities of less than 5 gallons may be flushed down drain. For larger quantities wipe up with an absorbent material or mop and dispose of in accordance with local, state and federal regulations. Dilute with water to minimize oxidizing effect on spilled surface.

VII Reactivity Data

Stable under normal use and storage conditions. Strong oxidizing agent. Reacts with other household chemicals such as toilet bowl cleaners, rust removers, vinegar, acids or ammonia containing products to produce hazardous gases, such as chlorine and other chlorinated species. Prolonged contact with metal may cause pitting or discoloration.

VIII Fire and Explosion Data

Not flammable or explosive. In a fire, cool containers to prevent rupture and release of sodium chlorate.

IX Physical Data

Boiling point----------------------212°F/100°C (decomposes)
Specific Gravity (H_2O=1)------------1.085
Solubility in Water-----------------complete
pH-----------------------------------11.4

Figure 6.22 Material Safety Data Sheet (MSDS)

Summary

For sensitive individuals, the presence of allergens in food is potentially life threatening. Currently, there is no cure for a food allergy. The only successful method to manage food allergy is avoidance of foods containing the allergen.

As described in the case study at the beginning of this chapter, product changeover provides an unintentional opportunity for a product that contains an allergen to contaminate a product that does not contain that particular allergen (i.e., walnuts). Equipment cleaning is a critical allergen control point for the production of a nonallergen-containing product following product changeover. All visible and non-visible product residues must be removed from food-contact surfaces by proper cleaning and sanitizing between products. The cleanup and sanitation steps used by bakery personnel should be verified and proven effective.

A good sanitation program starts with a neat, clean, and properly maintained building. A well-maintained facility will help protect food products from contamination, make proper stock rotation easier, prevent entrance of pests, reduce fire hazards, and contribute to the overall safety of the work environment. Proper cleaning and sanitizing also enhance the safety and quality of food and increase the life expectancy of equipment and facilities.

A retail food establishment's sanitation program consists of the standards, policies, and procedures that assure proper cleaning and sanitation. In-store sanitation standards must incorporate the requirements of government regulations and manufacturers' guidelines.

The effectiveness of a retail food establishment's sanitation program is directly related to the degree of commitment and concern demonstrated by top management. Managers must motivate their employees to follow prescribed sanitation practices and reward those employees who do so successfully.

Proper education and training is required to ensure proper sanitation practices are followed. Employees responsible for handling food, equipment, and cleanup of the building and grounds must be trained to accept sanitation as one of their key responsibilities. A clean, well-maintained facility will become a source of pride for employees.

For a sanitation program to be effective, cleaning tasks must be scheduled and individual employees assigned to complete the tasks.

Employees must be given instructions that include what they are to clean, how to clean it, and what tools and supplies are required to effectively clean it. Cleaning activities must be carried out by properly trained employees, and they must be closely monitored by supervisors to identify and correct problems when they occur.

A good sanitation program is a preventive program which anticipates and eliminates potential hazards before they become serious problems. The value of an effective program is difficult to measure in terms of dollars. Nonetheless, the value of effective sanitation can be evident in the form of diminished problems and increased goodwill with customers and the local regulatory community.

Discussion Questions (Short Answer)

1. What is the difference between cleaning and sanitizing?
2. Briefly discuss some of the factors that affect cleaning efficiency.
3. What are the main functions of detergents in the cleaning process?
4. What are the basic steps in the manual warewashing process?
5. Why must cleaned and sanitized equipment and utensils be air-dried?
6. How does water temperature affect chemical sanitizers?
7. What are some of the factors that affect the action of chemical sanitizers?
8. What is the best way to measure the strength of a chemical sanitizer?
9. What steps are commonly used to clean fixed equipment that is too large to be placed in a dishwashing machine or three-compartment sink?
10. What purposes do Material Safety Data Sheets serve in a retail food establishment?

Quiz 6 (Multiple Choice)

1. What are the proper steps in a manual warewashing operation after pre-flushing?
   a. Wash, rinse, sanitize, and towel dry.
   b. Rinse, wash, sanitize, and air-dry.
   c. Wash, rinse, sanitize, and air-dry.
   d. Rinse, wash, sanitize, and towel dry.
2. When sanitizing with hot water in a manual warewashing operation, the temperature of the water in the final rinse must be maintained at:
   a. 161°F (72°C).
   b. 171°F (77°C).
   c. 181° F (83°C).
   d. 191°F (88°C).
3. Which of the following statements is false?
   a. Dishes should be washed in very hot water (above 171°F, 77°C) to effectively remove soil from the surface.
   b. Pre-scraping helps to remove larger food particles from dishes, which helps keep the wash water clean.
   c. The cleaning compounds used in a retail food establishment must be tailored to the individual water supply.
   d. Cleaning is a process which removes soil and prevents accumulation of food residues on equipment, utensils, and surfaces.
4. The allowable range of water temperatures for chlorine sanitizing solutions is between ________________ and ________________.
   a. 55°F (13°C); 120°F (49°C).
   b. 75°F (24°C); 120°F (49°C).
   c. 95°F (35°C); 140°F (60°C).
   d. 120°F (49°C); 171°F (77°C).

5. The strength of the chemical sanitizer in the third compartment of the three-compartment sink must be checked frequently because:
   a. If the chemical is too strong, it may ruin fragile dishware.
   b. The chemical strength increases over time and may leave a toxic residue on the surface of equipment and utensils.
   c. The strength of chemical sanitizers may drop off as germs are killed off and the sanitizer is diluted with rinse water.
   d. The strength of the chemical increases as germs are killed off.
6. The first step in developing a cleaning program for environmental areas is to:
   a. Determine the cleaning needs.
   b. Assign cleaning jobs to employees.
   c. Obtain the equipment and supplies needed to implement the cleaning.
   d. Conduct a training program to teach employees how to clean properly.
7. Which of the following information is not provided by a Material Safety Data Sheet?
   a. Information about the physical and chemical characteristics of hazardous substances used in the facility.
   b. Information about how to use personal protective equipment and other devices to reduce a worker's risk of injury.
   c. Information about emergency procedures to use when exposed to a hazardous chemical.
   d. Information about approved hazardous waste sites that will accept unused portions of the hazardous materials.
8. Which of the following statements is false?
   a. Keeping things clean is the responsibility of every person working in the food industry.
   b. Cleanliness goes beyond the removal of visible soil.
   c. A good sanitation program, if properly organized, will function successfully without the support of top management.
   d. A good sanitation program starts with a neat, clean, and properly maintained building.

Answers to the multiple choice questions are provided in Appendix A.

References/Suggested Readings

Food and Drug Administration (2001). *2001 Food Code.* United States Public Health Service. Washington, DC.

Suggested Web Sites

All QA Products
www.allqa.com

Bowerman Associates
www.ebowerman.com

Champion Industries
www.championindustries.com

DiverseyLever Inc.
www.diverseylever.com

Duke Manufacturing Company
www.dukemfg.com

Ecolab
www.ecolab.com

Hobart Corporation
www.hobartcorp.com

Paper Thermometer Company
www.mv.com/ipusers/paperthermometer/

The Soap and Detergent Association
www.sdahq.org

Steritech
www.stertech.com

NSF International
www.nsf.org

Underwriters Laboratories, Inc.
www.ul.com

The Food Marketing Institute
www.fmi.org

Gateway to Government Food Safety Information
www.foodsafety.gov

United States Department of Agriculture (SDA)
www.usda.gov

United States Foods and Drug Administration (FDA)
www.fda.gov

7 Environmental Sanitation and Maintenance

Bulk Food Storage and Display Bins Serve more than Customers

The "Fresh-to-You" retail food establishment is known for its complete line of exotic ingredients, fresh produce, and hard-to-find items. It is very popular with shoppers and enjoys a high volume of business. It uses a lot of bin-type, self-service display and storage for rice, beans, peppers, and other dried foods. Customers scoop up the amounts of these products they want to buy and place them into plastic bags. The foods are frequently spilled as they are put into the plastic bags, and the areas around the bins are difficult to keep clean. There are bins of candy and dried fruits that are displayed and dispensed in the same manner in the aisle next to the dried vegetables.

The company hired to provide cleaning and general maintenance of the establishment reported to the manager they had seen signs of rodent and insect infestation. Rub marks had been seen along the walls behind the display bins and rodent droppings had been seen near the bins and in the storeroom. Cockroach egg cases had been found near the raisin and candy containers.

The manager called the pest control operator (PCO) who was under contract with the facility and asked for an immediate inspection. Following the inspection the PCO and the market manager held a meeting. What measures do you think they concluded should be taken to solve the pest problem at the establishment?

Learning Objectives

After reading this chapter, you should be able to:

- ▲ Describe how the visual appearance of a retail food establishment can affect a customer's opinion about the cleanliness and sanitation of the operation.
- ▲ Identify the criteria that should be used when selecting materials for floors, walls, and ceilings in food production, warewashing, and display areas of a retail food establishment.
- ▲ Identify the equipment and supplies that must be provided in a properly equipped restroom facility.
- ▲ Identify the main components of a properly equipped handwashing station and explain how conveniently located handwashing facilities contribute to good personal hygiene.
- ▲ Identify the types of plumbing hazards that can have a negative effect on public health and the devices commonly employed to prevent these hazards.
- ▲ Identify some types of pests that may be found in retail food establishments.
- ▲ Identify some common signs of pest infestation in retail food establishments.
- ▲ Describe how integrated pest management is used to control pests in retail food establishments.
- ▲ Explain how proper disposal and storage of garbage and refuse help prevent contamination and pest problems in a retail food establishment.

Essential Terms

Air gap	**Garbage**
Backflow	**Integrated Pest Management (IPM)**
Coving	**Refuse**
Cross connection	**Vacuum breaker**

Condition of the Establishment

A retail food establishment is made up of the building, its contents, and the surrounding land or property. The exterior of the establishment includes the outside of the building, parking areas, landscaping, doors and windows. To attract customers, the exterior of the building should be attractive and make a good impression.

The exterior of the facility must be free of litter and debris that could attract and harbor pests and ruin the appearance of the facility. Grass and weeds should be regularly mowed to eliminate harborage areas for insects, rodents, and other pests. Walking and driving surfaces should be constructed of concrete, asphalt, gravel, or similar materials to facilitate maintenance and control dust. These surfaces should also be properly graded to prevent rainwater from pooling and standing on parking lots and sidewalks.

Proper Water Supply and Sewage Disposal System

An adequate water supply and proper sewage disposal are vital to the sanitation of retail food establishments. The water source and ability to meet hot water generation needs should be sufficient to meet demands of the food establishment. Drinking water for retail food establishments must be obtained from an approved source. Most establishments will be connected to a public water system. However, when a private well or other nonpublic water system is used, it must be constructed, maintained, and operated according to the water quality requirements of the jurisdiction. Wells must be located and constructed in a manner that will protect them from sewage and other sources of contamination. According to the *FDA Food Code*, water from a nonpublic water system must be sampled and tested at least annually and as required by state water quality regulations. Periodic sampling is required to monitor the safety of the water and to detect any change in water quality.

To render the water safe, a drinking water system must be flushed and disinfected before being placed into service. The *FDA Food Code* prescribes a drinking water system is flushed after construction, repair, or modification and after an emergency situation, such as flood, that may introduce contaminants to the system.

A reservoir that is used to supply water to a device such as a produce fogger must be maintained in accordance with the manufacturer's specifications. The reservoir must be cleaned at least once a week in accordance with manufacturer's specifications or according to the following procedures, whichever is more stringent:

1. Drain and completely disassemble the water and aerosol contact parts.
2. Brush-clean the reservoir, aerosol tubing, and discharge nozzles with a suitable detergent solution.
3. Flush the complete system with water to remove the detergent solution and particulate accumulation.
4. Rinse by immersing, spraying, or swabbing the reservoir, aerosol tubing, and discharge nozzles with at least 50 ppm hypochlorite solution.

The use of nonpotable water sources in retail food establishments must be approved by the regulatory authority in the jurisdiction. Nonpotable water sources may only be used for non-culinary purposes such as air handling systems, cooling systems, and fire protection.

Proper disposal of sewage greatly reduces the risk of fecal contamination of food and water. The *FDA Food Code* requires sewage from retail food establishments to be disposed through an approved facility that is:

- ▲ A public sewage treatment plant, or
- ▲ An individual sewage disposal system that is sized, constructed, maintained, and operated according to rules and regulations of the jurisdiction.

Condition of Building

The entrance area to a retail food establishment creates a lasting impression. The cleanliness and attractiveness customers view upon entering the building influence their overall shopping experience (Figure 7.1). The entrance should be easy to find and have a pleasing appearance. The entrances to the building should provide easy access from streets and parking areas. Entrance doors should be self-closing to discourage flying insects. Clearly mark the doors for entrance and exit to prevent accidents between those entering and leaving the establishment.

Figure 7.1 Keep the Outside of the Establishment Clean

Floors, Walls, and Ceilings

Proper construction, repair, and cleaning of floors, walls, and ceilings are important elements of an effective sanitation program. Consider the specific needs of the different departments when selecting materials for floors, walls and ceilings. Some criteria that should be used for all departments are:

- ▲ Sanitation
- ▲ Safety
- ▲ Durability
- ▲ Comfort
- ▲ Cost.

Smooth and easy-to-clean surfaces are needed in food preparation areas, store rooms (including dry storage areas and walk-in refrigerators), and warewashing areas.

For floor coverings, walls, wall coverings, and ceilings, use nonabsorbent materials free from cracks or crevices where soil may lodge, and from imperfections that might cause accidents. Choose surfaces resistant to damage and deterioration from water, cleaning agents, and the repeated scrubbing used to keep them in good condition.

Floors

The preferred floor materials in food preparation and warewashing areas include:

- ▲ Terrazzo
- ▲ Quarry tile
- ▲ Asphalt tile
- ▲ Ceramic tile
- ▲ Concrete sealed with epoxy or a similar material to make it durable and nonabsorbent.

Avoid using wood and vinyl floor covering in food production and warewashing areas. These materials are generally not durable and easy to clean. They also absorb water, soil, and other forms of contamination that can cause sanitation problems when used in these areas.

The *FDA Food Code* prohibits the use of carpeting in:

- ▲ Food preparation areas
- ▲ Walk-in refrigerators
- ▲ Warewashing areas
- ▲ Toilet room areas where handwashing lavatories, toilets, and urinals are located
- ▲ Refuse storage rooms or other areas subject to moisture.

Floors graded to drains are needed in retail food establishments where water-flush methods are use for cleaning. In addition, the floor and wall must be coved and sealed. **Coving** is a curved sealed edge between the floor and wall that eliminates sharp corners or gaps that would make cleaning difficult and ineffective (Figure 7.2). When cleaning methods other than water flushing are used for cleaning floors, the floor and wall juncture must be coved with a gap of no more than $^1/_{32}$ inch (1 mm.) between the floor and wall.

Figure 7.2 Coving

Slips and falls are the most common types of accidents in retail food establishments. In many instances, slippery floors cause accidents that result in personal injury and broken equipment. Use mats and other forms of anti-slip floor coverings where necessary to protect the safety of workers (Figure 7.3). These devices should also be impervious, nonabsorbent, and easy to clean. Good sanitation will reduce the risk of slips and falls by removing water and food wastes from floor areas.

The *FDA Food Code* prohibits the use of sawdust, diatomaceous earth, or similar materials on floors except where they are applied in small amounts to collect liquid spills or drippage immediately before spot cleaning.

Walls and Ceilings

Smooth, nonabsorbent, and easy-to-clean walls and ceilings must be provided in food production and warewashing areas, walk-in refrigerators, and toilet facilities. Light colors enhance the artificial lighting in these areas and make soil easy to see for better cleaning. In areas cleaned frequently, use walls and wall coverings constructed of materials such as ceramic tile, stainless steel, or fiberglass. Concrete, porous blocks, and bricks should only be used in dry storage areas unless they are finished and sealed to provide a smooth, nonabsorbent, and easily cleanable surface.

Figure 7.3 Anti-Slip Mat in Front of a Three-Compartment Sink

Ceilings should be constructed of nonporous, easily cleanable materials. Studs, rafters, joists, or pipes must not be exposed in walk-in refrigeration units, food preparation and warewashing areas, and toilet rooms. Light fixtures, ventilation system components, wall-mounted fans, decorative items, and other attachments to walls and ceilings must be easy to clean and maintained in good repair.

Restroom Sanitation

Toilet facilities are required for all employees. Employee restrooms must be conveniently located and accessible to employees during all hours of operation. Toilet facilities near work areas promote good personal hygiene, reduce lost productivity, and permit closer supervision of employees. Toilet rooms in retail food establishments must be completely enclosed and provided with tight-fitting and self-closing doors.

Materials used in the construction of toilet rooms and toilet fixtures must be durable and easily cleanable. The floors, walls and fixtures in toilet areas must be clean and well-maintained (see Figure 7.4). Supply toilet tissue at each toilet. Provide easy-to-clean containers for waste materials and have at least one covered container in toilet rooms used by women.

Figure 7.4 Restroom Sanitation

Poor sanitation in toilet areas can spread disease. Dirty toilet facilities can also have a negative effect on the attitudes and work habits of the establishment's employees. Include these areas in the routine cleaning program to ensure they are kept clean and in good repair. Never store food or items associated with food handling in restroom areas.

Handwashing Facilities

In earlier chapters of this book, you learned about the importance of food employees' hands as sources of contamination and cross contamination. Food employees must know when and how to wash their hands. Conveniently located and properly equipped handwashing facilities are key factors in getting employees to wash their hands. Handwashing stations (Figure 7.5) should be located in or near areas where food is prepared or handled and in warewashing areas. Handwashing stations must also be located in or adjacent to restrooms.

The number of handwashing stations required and their installation are usually set by local health or plumbing codes.

Figure 7.5 Handwashing Station

A handwashing station must be equipped with hot and cold running water under pressure, a supply of soap, and a means to dry hands. A handwashing lavatory must be able to provide water at a temperature of at least 100°F (38°C) through a mixing valve or combination faucet. If a self-closing, slow-closing, or metering faucet is used, it must provide a flow of water for at least 15 seconds without the need to be reactivated.

Each handwashing station must be equipped with a dispenser containing liquid or powdered soap. The use of bar soap is frequently discouraged by regulatory agencies because bar soap can become contaminated with germs and soil.

Individual disposable towels and mechanical hot-air dryers are the preferred hand-drying devices. Most local health departments do not recommend retractable cloth towel dispenser systems because there are too many possibilities for contamination. Common cloth towels, used multiple times by employees to dry their hands, are also prohibited.

Keep handwashing stations clean and well-maintained and never use them for purposes other than handwashing. It is equally important to remember sinks used for preparing produce and warewashing must not be used for handwashing.

Plumbing Hazards in Retail Food Establishments

The *FDA Food Code* includes many different components within the definition of a plumbing system. In particular, it identifies the

- Water supply and distribution pipes
- Plumbing fixtures and traps
- Soil, waste, and vent pipes
- Sanitary and storm sewers
- Building drains, including their respective connections and devices within the building and at the site.

A properly designed and installed plumbing system is extremely important to food sanitation. Contamination of the public water supply in a retail food establishment can create a serious public health hazard. Numerous outbreaks of gastroenteritis, dysentery, typhoid fever, and chemical poisonings have been traced to cross connections and other types of plumbing hazards in retail food establishments. The plumbing system in a retail food establishment must be maintained in good repair. When repairs to the plumbing system are required, they must be completed in accordance with applicable local ordinances and state regulations.

Cross Connections

A **cross connection** is any physical link through which contaminants from drains, sewers, or waste pipes can enter a potable (safe to drink) water supply. A cross connection may be either direct or indirect. A direct cross connection occurs when a potable water system is directly connected to a drain, sewer, nonpotable water supply, or other source of contamination. An indirect cross connection is where the source of contamination (sewage, chemicals, etc.) may be blown across, sucked into, or diverted into a safe water supply.

One of the most commonly found cross connections in retail food establishments is a hose attached to a service sink, with the end of the hose submerged in soapy water that is used to clean floors and other non-food-contact surfaces.

Backflow

Backflow is the backward flow of contaminated water into a potable water supply. Backflow occurs most frequently under two conditions:

- ▲ Backpressure where contamination is forced into a potable water system through a connection that has a higher pressure than the water system
- ▲ Backsiphonage when there is reduced pressure or a vacuum formed in the water system (Figure 7.6). This might be caused by a water main break, the shutdown of a portion of the system for repairs, or heavy water use during a fire.

Figure 7.6 Backsiphonage

Methods and Devices to Prevent Backflow

The plumbing system in a retail food establishment must be designed, constructed, and installed according to the plumbing code in the local jurisdiction. A properly designed and installed plumbing system will keep food, equipment, and utensils from becoming contaminated with disease-causing microorganisms found in sewage and other pollutants. Cross connections and backflow can be prevented by using the following devices:

- Air gap
- Atmospheric vacuum breakers
- Pressure type vacuum breakers
- Double check valves
- Reduced pressure principle backflow preventers.

Air Gap

An **air gap** is the physical separation of the potable and nonpotable system by a vertical air space. The vertical distance between the supply pipe and the flood rim must be at least two times the diameter of the supply pipe (2D), but never less than 1 inch (25 mm.). An air gap is the most dependable backflow prevention device. Figure 7.7 shows how an air gap may be used between a faucet and the rim of a sink and between a drain line and the establishment's sewer system.

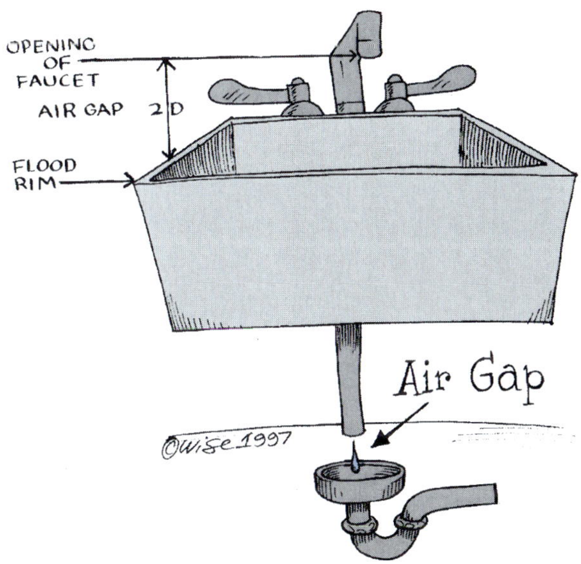

Figure 7.7 Air Gap

Atmospheric Type Vacuum Breaker

A **vacuum breaker** prevents backflow to a water supply by venting water to atmosphere when backflow conditions exist. Atmospheric type anti-siphon vacuum breakers are commonly used in retail food establishments. These devices use an atmospheric vent in combination with a check valve (Figure 7.8). The valve's air inlet closes when the potable water flows in the normal direction. But, as water ceases to flow, the air inlet opens, thus interrupting the possible backsiphonage effect.

Figure 7.8 Atmospheric Type Vacuum Breakers

Atmospheric vacuum breakers can be used on most inlet type water connections that are not subject to backpressure. This includes low inlet feeds to receptacles containing toxic and nontoxic substances.

Pressure Type Vacuum Breakers

Pressure type **vacuum breakers** are designed for use under continuous supply pressure, but are not good under backpressure (Figure 7.9). The device is usually spring loaded and designed to operate after extended periods of hydrostatic pressure. These devices should only be used where nonpressure vacuum breakers cannot be used.

Figure 7.9 Pressure Type Vacuum Breaker

Double Check Valve

A *double check valve* consists of two internally loaded, independently operating check valves, together with tightly closing resilient seated shut-off values upstream and downstream of the check valves (Figure 7.10).

A double check valve assembly may be used as protection for all direct connections through which foreign materials might enter the potable water system.

Figure 7.10 Double Check Valve

Reduced Pressure Principle Backflow Preventers

This assembly consists of two internally loaded independently operating check valves and a mechanically independent, hydraulically dependent relief valve located between the check valves (Figure 7.11).

Figure 7.11 Reduced Pressure Backflow Preventer

Reduced pressure zone devices may be used on all direct connections that may be subject to backpressure or backsiphonage. Use this device when there is the possibility of contamination by material that is a potential health hazard.

These devices are commonly installed as mainline protection to protect the municipal water supply. They should also be used on branch line applications where nonpotable fluid would constitute a health hazard if introduced into the potable water supply system.

Backflow Prevention Devices on Carbonators

Carbonators on soft drink dispensers form carbonic acid by mixing carbon dioxide with water. The carbonic acid is then mixed with the syrups to produce the soft drinks. If the carbon dioxide backs up into a copper water line, the carbonic acid will dissolve some of the copper. The water containing the dissolved copper will then be used in dispensing soft drinks, and the first few customers receiving the drinks are likely to suffer the symptoms of copper poisoning. An air gap or a vented backflow prevention device meeting American Society of Sanitary Engineering (ASSE) Standard No. 1022 must be installed upstream from a carbonating device and downstream from any copper in the water supply line to reduce incidences of copper poisoning (Figure 7.12).

Figure 7.12 Backflow Preventer

Grease Traps

Retail food establishments that produce large amounts of grease from a meat department and similar operations should be equipped with a grease trap (Figure 7.13). These devices remove liquid grease and fats after they have hardened and become separated from the wastewater. Grease traps are especially important when the retail food establishment is connected to a septic system or other type of on-site wastewater treatment and disposal system. A grease trap must be located for easily accessible cleaning. Failure to locate a grease trap so it can be properly cleaned and maintained can result in the harborage of pests and/or the failure of the facility's sewage system.

Figure 7.13 Grease Trap Suspended from Basement Ceiling

Garbage and Refuse Sanitation

Garbage is the term applied to food wastes that cannot be recycled. **Refuse** is trash, rubbish and other types of solid waste that are not disposed of through the sewage system. Proper storage and disposal of garbage and refuse are required at retail food establishments to protect food and equipment from contamination. Insects, rodents, and other pests are also less likely to be attracted to these establishments when garbage and refuse are properly managed. Effective waste management requires:

- Proper handling and short-term storage of the materials inside the operation
- Proper storage of the waste outside the building until it is picked up by a commercial refuse disposal company.

Inside Storage

Waste containers must be provided in all departments and other areas in a retail food establishment where refuse is produced or discarded. Containers used to collect garbage and refuse must be:

- ▲ Durable
- ▲ Cleanable
- ▲ Insect and rodent-proof
- ▲ Leak-proof
- ▲ Nonabsorbent
- ▲ Covered with a tight-fitting lid when not in use.

Plastic bags and wet strength paper bags are frequently used to line waste containers (Figure 7.14). These bags can greatly reduce the amount of cleaning needed to maintain the cans in good condition, and they make the waste easier to handle. Do not place waste containers in locations where they might create a public health nuisance or interfere with cleaning environmental areas.

Figure 7.14 Waste Storage Container with Plastic Liner

Storage rooms are commonly provided for holding recyclable materials and returnable items in retail food establishments. These areas must be large enough to hold all materials awaiting pickup. Shelving and other equipment used inside these storage rooms must be designed and constructed for easy cleaning.

Refuse storage rooms and containers must be cleaned as part of the establishment's routine cleaning program. Remember, dirty equipment and containers attract insects and rodents. Keeping the area clean is your best defense against pests. Provide suitable cleaning equipment and supplies such as high pressure pumps, hot water, steam, and detergent to thoroughly clean equipment and receptacles. When cleaning this equipment, be careful not to contaminate food, equipment, utensils, linens, or single-service and single-use articles. Waste water produced while cleaning the equipment and receptacles is considered to be sewage. It must be disposed of through an approved sanitary sewage system or other system constructed, maintained, and operated according to law.

Outside Storage

A retail food establishment should also have an outside storage area and enclosure (Figure 7.15) to hold refuse, recyclables, and returnables awaiting pickup. An outdoor storage surface should be durable, cleanable, and maintained in good repair. Dumpsters and storage areas must be covered with tight-fitting lids, doors, or covers to discourage insects, rodents, and other types of pests.

Figure 7.15 Outside Garbage and Refuse Station

Refuse and garbage should be removed from the site as often as necessary to prevent objectionable odors and avoid conditions that attract or harbor insects and rodents.

Outdoor storage areas must be kept clean and free of litter. Suitable cleaning equipment and supplies must be available to clean the equipment and receptacles. Refuse storage equipment and receptacles must have drains and drain plugs must be in place.

Compactors and other equipment for refuse, recyclables, and returnables must be installed to minimize the accumulation of debris (Figure 7.16). Even a thin layer of food or grease left on the inside or outside of a garbage container is sufficient to support a population of annoying fruit flies. It is best, when possible, to use leak-proof containers. Always make sure you clean under and around these units to prevent insect and rodent harborage.

Figure 7.16 Trash Bailer

Cardboard or other packaging material (that does not contain food residues) may be stored outside for regularly scheduled pickup for recycling or disposal. If materials are stored so they do not create a rodent-harborage problem, a covered receptacle is not required.

Some retail food establishments may provide redeeming machines for recyclables or returnables. According to the *FDA Food Code*, a redeeming machine may be located in the packaged food storage area or consumer area of a retail food establishment if food, equipment, utensils and linens, and single-service and single-use articles are not subject to contamination from the machine and a public health nuisance is not created.

Pest Control

Every retail food establishment should have a pest control program. The targets of this program are insects and rodents that can spread disease and damage food. These pests carry disease-causing microorganisms in and on their bodies and can transfer them to food and food-contact surfaces. Pests also destroy millions of dollars of food each year by eating it or by contaminating it with urine and feces. The pests that are the most common targets of pest control programs include rats, mice, house flies, cockroaches, small moths, and beetles.

The benefits of proper cleaning and sanitizing, time and temperature controls, good personal hygiene and cross contamination control can be wasted if insects and rodents are allowed to contaminate foods and food-contact surfaces. Any retail food establishment can have an occasional insect or rodent problem. However, their continual presence causes major problems and indicates a lack of good sanitation and control measures.

The key element of a successful pest control program is prevention (Figure 7.17). However, no single measure will effectively prevent or control insects and rodents in retail food establishments. It takes a combination of three separate activities to keep pests in check. You must:

- ▲ Prevent entry of insects and rodents into the establishment
- ▲ Eliminate food, water, and places where insects and rodents can hide
- ▲ Implement an integrated pest management (IPM) program to control insect and rodent pests that enter the establishment.

Figure 7.17 Prevent Pests from Invading Your Establishment

Insects

What insects lack in size, they more than make up for in numbers. Insects may spread diseases, contaminate food, destroy property, or be nuisances in retail food establishments. Insects need water, food, and a breeding place in order to survive. The best method of insect control is keeping them out of the establishment coupled with good sanitation and integrated pest management when needed.

Flies

Houseflies, blowflies, and fruit flies are the types of flies most commonly found in retail food establishments.

The housefly is the one most likely to spread disease. Twenty-one species of flies are categorized as "disease-causing flies" because they are proven carriers of *E. coli, Salmonella, Shigella* and other bacteria that can cause foodborne illness (Olsen, 1998). When a fly lands on filth, such as human and animal feces or garbage, some of the material, including bacteria, sticks to the fly's body and the hairs on its legs. When the fly goes into areas where food is prepared or eaten, it walks on food and food-contact surfaces. The filth from the fly's body and legs is deposited there and contaminates the food and surfaces (Figure 7.18). When a fly feeds on waste material, it takes bacteria into its body. The housefly cannot chew solid food. Instead, it vomits on the food to soften it before eating it. When the fly vomits, some of the bacteria in its stomach are deposited onto food and food-contact surfaces causing them to become contaminated.

Figure 7.18 Flies Spread Disease

Blowflies are usually larger than houseflies and are a shiny blue, green, or bronze color. Blowflies have a keen sense of smell and are attracted by the odors produced by food establishments and food-processing plants.

Fruit flies are the smallest of the three flies and are attracted by decaying fruit and other organic materials. Fruit flies are known to spread plant diseases. Their role in the spread of microorganisms that cause infections in humans is currently being investigated by researchers at the USDA.

Fly Control

Eliminating the insect's food supply is the first step in fly control. Food should be protected from flies during preparation, storage and display. In many cases, improperly stored trash and garbage is the main source of food for flies. It is important to store trash, garbage and other waste in containers and storage rooms that are fly-tight. These garbage storage facilities must also be cleaned on a regular basis to keep them from becoming an attractive nuisance for flies. Effective control also requires regular cleaning of food production, storage and display areas as well as toilet facilities.

Windows and service doors must be equipped with screens to prevent flies from entering the retail food establishment. Entrances and openings at loading docks must be equipped with air curtains (Figure 7.19) to prevent flies and other flying insects from entering the building.

Figure 7.19 Air Curtain

Flying insect trapping systems are commonly used to control flying insects (Figure 7.20). These systems include electrocuting and non-electrocuting type devices. Electrocuting devices contain a light source that attracts the insects to a high voltage wire grid. Upon contact with the grids, the high voltage electrocutes the insects. Recent evidence suggests insect fragments are produced when the insects are electrocuted. These fragments can become a source of contamination for food and food-contact surfaces. The *FDA Food Code* prohibits insect electrocutor traps from being installed over a food preparation area. When used in retail food establishments, these devices must be installed so dead insects and insect parts cannot fall onto food; clean equipment, utensils, and linens; and unwrapped single-service and single-use articles. There are also phosphorescent fly lights that make flies unconscious and trap them on a piece of sticky paper located at the bottom of the trap.

Figure 7.20 Flying Insect Electrocutor

Non-electrocuting systems operate with glue traps and pheromone attractants. They are similar in design to the electrocuting devices but instead of an electrocuting grid, a glue board is used to capture the flying insect. These devices are very effective and are allowed in areas where electrocuting devices are not.

Chemical insecticides should be applied by a professional pest control operator as a last resort. Remember, insecticides are a supplement to, never a substitute for, proper food handling and a clean facility.

Cockroaches

American, Oriental, German, and Brown Banded are the four most common cockroaches in the United States. Cockroaches carry bacteria and other organisms that can cause human illness. Roaches live in sewers and other areas where filth is found. They carry bacteria on their hairy legs and body, as well as in their intestinal tract. They can contaminate food and food-contact surfaces when they crawl around food storage, preparation and display areas.

The German cockroach is the one most frequently found in retail food establishments (Figure 7.21). The German cockroach prefers a warm, dark and moist environment. This type of cockroach is frequently found in and around hot water heaters, ovens, steam tables, and dishwashing machines. The female German cockroach carries 24 to 48 eggs with her in an egg capsule until they are ready to hatch. This is one reason why the German cockroach is the most successful of the roaches.

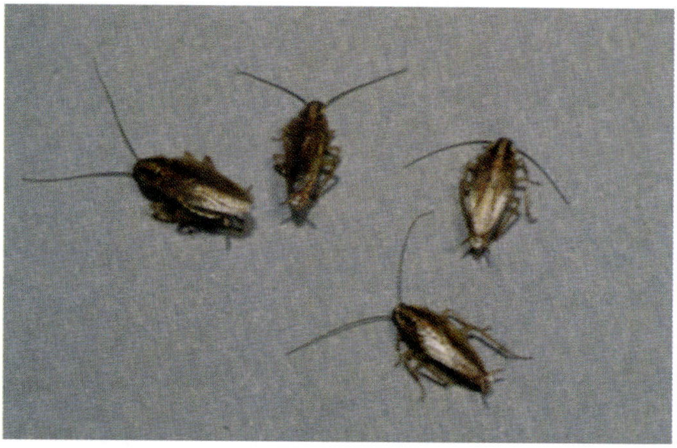

Figure 7.21 German Cockroaches

Roaches are nocturnal creatures that avoid light. Roaches hide in cracks and crevices under and behind equipment and in toilet rooms and other areas of the facility. It is possible to have a serious cockroach infestation and not realize it until they are caught by surprise some night when a light is turned on suddenly in a room.

Cockroach Control

Cockroach control requires good housekeeping both indoors and out. Eliminate hiding places by picking up unwanted materials, such as boxes and rags. Fill cracks and crevices in floors and walls and around equipment. Doors and windows should be tight-fitting and kept in a closed position. If windows and doors are kept open for ventilation or other purposes, they must be properly screened or protected by air curtains or other means.

Cockroaches frequently enter retail food establishments with incoming products and supplies. Check incoming shipments for signs of infestation such as egg cases and live roaches. Store food in containers that are insect proof and have tight-fitting lids. Keep floors, walls and equipment clean and free of food wastes. Frequent cleaning will help remove egg cases and reduce the roach population.

Chemical control of cockroaches is only recommended when used in combination with other control methods. Insecticides should be applied by a professional pest control operator. These individuals know how to handle insecticides safely to avoid contaminating food and food-contact surfaces.

Effective cockroach control requires the use of materials suitable for the area of the establishment you are working in. It is often necessary to use more than one type of product or procedure to eliminate an infestation. Some treatments call for the use of residual insecticides that remain active for several days after they are applied. The life expectancy of a residual insecticide will depend on the type of material treated and the level of sanitation and housekeeping in the establishment. Make certain the pest control operator has experience in treating retail food establishments and pays attention to areas such as door trim, under and around equipment, windows, baseboards, and wall-mounted equipment and dispensers. A second treatment may be required in facilities experiencing major infestations. This second application will be applied within 10 to 14 days after the first and is necessary to disrupt the breeding cycle of the cockroach.

Commercially available baits can also be used throughout an establishment as well as in "blind" areas such as false ceilings, wall voids, or even electric control panels and junction boxes. Baits come in granular, gel, and paste formulations as well as in self-contained bait stations. Proper placement of the bait is critical to the success of a baiting program.

Moths and Beetles

Moths and beetles cause concern since they invade certain foods and can do extensive damage. The Indian Meal Moth, saw-toothed grain beetle, flour weevil, and rice weevil are examples of insects that can be found in stored products in retail food establishments (Figure 7.22). These insects create problems of wasted food and nuisance rather than disease.

The Indian Meal Moth is one of the more common moths infesting stored grains and grain products. Although manufacturers attempt to deliver food that is virtually pest free, it is often difficult to achieve that level of success.

Figure 7.22 (a) Indian Meal Moth and (b) Saw-Toothed Grain Beetle

The life cycle of the Indian Meal Moth includes egg, larva, pupa and adult. The larva feeds on coarse grains (oatmeal and cornmeal), pasta, cereals, spices, raisins and other dried fruits, nuts, and dry pet foods. The larva does damage as it feeds on food, spinning silk as it goes and leaving a mat of webbing and feces on the food's surface. The extensive webbing can spoil more food than the larva can consume.

The first step in managing an Indian Meal Moth problem is inspecting for and then removing infested food products. Once the suspect food or foods have been identified and removed, thoroughly vacuum the area and wash equipment and utensils with soap and water. Special attention should be paid to cracks, crevices, corners, and under pallets where food may have accumulated. In many cases, thorough cleanup will control these insects.

A residual insecticide may be applied in cracks, crevices and corners to help eliminate the larvae of this pest from treated areas. When insecticide is used, great care must be taken not to contaminate food. Pheromone traps are now available for this pest. The trap contains a chemical lure that gives off the same odor as a female moth. Males are attracted to the trap and are caught on a sticky surface. These traps provide an effective means to monitor for the adult moth to determine if there is an ongoing infestation.

Other Controls for Moths and Beetles in Stored Foods

Examine incoming shipments for signs of infestation and keep infested products away from other stock until they are ready for disposal. Control of moths and beetles is aided by FIFO (first-in, first-out) stock rotation. Keep dry food storage areas cool as cool temperatures limit growth of these insects and reduce the rate of egg laying. All opened packages or sacks should be either used immediately or stored in covered containers.

Rodents

Rodents adapt easily to human environments and tolerate a wide range of conditions. They are known to carry microorganisms that can cause a number of human diseases including salmonellosis, plague, and murine typhus. Rodents also consume and damage large quantities of foods each year. Rats typically carry their food back to their nest rather than eat it where it is found. Rats are very shy creatures and are seldom seen in a retail food establishment.

Domestic rodents in the United States include the *Norway rat*, the *roof rat*, and the *house mouse* (Figure 7.23). The Norway rat is the one most commonly found in the United States. Other common names for the Norway rat are brown rat, sewer rat, and wharf rat.

Figure 7.23 House Mouse

(Courtesy of Bell Laboratories, Inc.)

The *Norway rat* is primarily a burrowing rat. It hides in burrows in the ground and around buildings and in sewers. Norway rats will eat almost any food but prefer garbage, meat, fish, and cereal. They stay close to food and water, and their range of travel is usually no more than 100 to 150 feet.

The *roof rat* is smaller than the Norway rat but is a very agile climber. It generally harbors in the upper floors of buildings but is sometimes found in sewers. Roof rats prefer vegetables, fruits, cereal and grain for food. The range of travel for the roof rat is also about 100 to 150 feet.

The *house mouse* is the smallest of the domestic rodents. It is found primarily in and around buildings, nesting in walls, cabinets, and stored goods. The house mouse is a nibbler, and it prefers cereal and grain. Its range of travel is 10 to 30 feet. Mice cause more economic damage than rats. This is due in part to the fact people are more tolerant of mice than rats.

Signs of Rodent Infestation

Rats and mice are nocturnal. Therefore, it is unusual to see them during the daytime. It is necessary to look for signs of rodent activity to determine if there is an infestation problem. When looking for signs of rodent activity, it is important to determine the type of rodent, whether it is a new or old problem and whether there is a light or heavy infestation.

Droppings

The presence of rat or mouse feces (Figure 7.24) is one of the best indications of an infestation. Fresh droppings are usually moist, soft and shiny, whereas old droppings become dry and hard. Norway rat droppings are the largest and have rounded ends. They look a lot like black jelly beans. Roof rat droppings are smaller and more regular in form. The droppings of the house mouse are very small and pointed at each end. They look something like dark grains of rice.

Figure 7.24 Rodent Droppings

Runways and Burrows

Rats are very cautious and repeatedly use the same paths and trails. Outdoors in grass and weeds, you may see paths worn down from repeated activity and are 2 to 3 inches wide.

The Norway rat prefers to burrow for nesting and harborage. Burrows are found in earth banks, along walls, and under rubbish. Rat holes are about 3 inches in diameter whereas mouse holes are only about 1 inch in diameter. If a burrow is active, it will be free of cobwebs and dust. The presence of fresh food or freshly dug earth at the entrance of the burrow also indicates an active burrow.

Rub marks

Rats prefer to stay close to walls where they can keep their highly sensitive whiskers in contact with the wall. As a rat runs along a wall, its body rubs against the wall or baseboard. The oil and filth from the rat's body are deposited on the wall and create a black mark called a "rub mark." Mice do not leave rub marks that are detectable, except when the infestation is especially heavy.

Gnawings

The incisor teeth of rats grow 4 to 6 inches a year. As a result, rats have to keep these teeth filed down in order to keep them short enough to use. Gnawings in wood are fresh if they are light colored and show well-defined teeth marks.

Tracks

Tracks may be observed along rat or mouse runs both indoors and outdoors. Look for tracks in dust in little-used rooms and in mud around puddles. Rat tracks may be 1½ inches long.

Miscellaneous Signs

Rodent urine stains can be seen with an ultraviolet light (black light). Rats leave a different pattern than mice. Rat and mouse hairs may be found along walls, etc. When examined under a microscope, they can be distinguished from other animal hairs.

Rodent Control

Effective rodent control begins with a building and grounds that will not provide a source of food, shelter, and breeding areas. The grounds around the food establishment should be free of litter, waste, refuse, uncut weeds, and grass. Unused equipment, boxes, crates, pallets and other materials should be neatly stored to eliminate places where pests might hide.

Get rid of all unwanted materials that may provide food and shelter for rodents. This means storing trash and garbage in approved-type containers with tight-fitting lids. Plastic liners help to keep cans clean, but they will not exclude rats and mice. Garbage must be removed frequently from the retail food establishment to cut off the food supply. Proper storage of returnable items is also necessary to avoid attracting rodents.

Rats can enter a building through holes as small as ½ inch. Mice can enter through even smaller holes. Buildings and foundations should be constructed to prevent rodent entry, and all entrances and loading and unloading areas should be equipped with self-closing doors and door flashings that will serve as suitable rodent barriers (Figure 7.25). Metal screens with holes no larger than ¼ inch should be installed over all floor drains.

Figure 7.25 Keep the Building Sealed Tight to Prevent Rodent Entry

Traps are useful around retail food establishments where rodenticides are not permitted or are hazardous. Live traps can be used for collecting live rats. Check traps at least once every 24 hours. Killer or snap traps can also be used as part of a rodent-control program. When using these types of traps, place them at right angles to the wall along rodent runways with the trigger side closest to the wall. Glueboards are shallow trays that have a very sticky surface. The mouse's feet stick to the board when they walk on it, and it is caught. Glueboards should be placed next to and running parallel with the wall.

Rodenticides are hazardous chemicals that can contaminate food and food-contact surfaces if not handled properly. Baits should be used outdoors to stop rodents at the outer boundaries of your property. These substances should be applied by a professional pest control operator. Baits should be placed in a tamper-proof, locked bait box that will prevent children and pets from being exposed to the toxic chemicals inside (Figure 7.26). Always make certain pesticides are stored in properly labeled containers, away from food in a secure place. Dispose of containers safely and know emergency measures for treating accidental poisoning.

The use of tracking powder pesticides is prohibited in retail food establishments. These types of pesticides can be dispersed throughout the establishment and directly or indirectly contaminate food, equipment, utensils, linens, and single-service/single-use articles. This contamination could adversely affect both the safety of the food and the general environment.

a. Bait Box b. Bait Box c. Glueboard

Figure 7.26 Bait Boxes and Glueboard
(Courtesy of J.T. Eaton & Company, Inc.)

Integrated Pest Management (IPM)

Modern pest control operators use integrated pest management as the primary method to control pests in retail food establishments. **Integrated pest management** is a system that uses a combination of sanitation, mechanical, and chemical procedures to control pests. Chemical pesticides are used only as a last resort and only in the amount needed to support the other control measures in the IPM program. If chemical pesticides are used, they must be approved for use as specified in the Code of Federal Regulations (40 CFR 152).

The National Pest Management Association (NPMA) recommends a five-step program for IPM:

1. Inspection.
2. Identification.
3. Sanitation.
4. Application of two or more pest management procedures.
5. Evaluation of effectiveness through follow-up inspections.

There are many benefits produced by using an integrated pest management program. An IPM program is more efficient and cost effective than programs that rely exclusively on chemicals to control pests. IPM is also longer lasting and safer for you, your employees, and your customers.

Summary

The case presented at the beginning of the chapter illustrates how pests can pose a serious challenge for retail food establishments. Pests are known to contaminate food and food-contact surfaces, and they can carry disease-causing microorganisms that are harmful to humans. Pests are attracted to retail food establishments by the food, water and pleasant odors they find there. In order to keep pests under control, retail food establishments must prevent them from entering the facility, eliminate sources of food and shelter, and implement an effective sanitation program. Without these prerequisite activities in place, it would be impossible for the Pest Control Operator to do his job properly.

Surveys of customers show cleanliness is a top consideration when choosing a place to eat or shop for food. Customer satisfaction is highest in retail food establishments that are clean and bright and where quality food products are safely handled and displayed.

Proper construction and repair and cleaning of floors, walls, and ceilings are important parts of an effective sanitation program. Sanitation, safety, durability, comfort, and cost are the main criteria you will use when selecting materials for floors and walls. Surfaces of floors and walls should be resistant to damage and deterioration from the water, detergents, and repeated scrubbings used to keep them clean. Walls and ceilings should be light colored to show soil and enhance the artificial lighting used in food preparation, handling and display areas.

Good personal hygiene, which includes proper handwashing, is an important factor in preventing the spread of foodborne disease-causing microorganisms. Handwashing stations must be conveniently located to enable food handlers to wash their hands as necessary throughout the workday.

Clean and suitably equipped toilet facilities must be provided for employees. These facilities must be kept clean and in good repair to prevent the spread of disease and promote good personal hygiene.

A properly designed, constructed, and installed plumbing system is very important to food sanitation. Numerous outbreaks of gastroenteritis, dysentery, typhoid fever, and chemical poisonings have been traced to cross connections and other types of plumbing hazards in retail food establishments. Air gaps and mechanical vacuum breakers are used to protect the municipal water supply. Consult a professional plumber or your local plumbing code for details about the plumbing requirements in your jurisdiction.

Proper storage and disposal of garbage and refuse are necessary to prevent contamination of food and equipment and avoid attracting insects, rodents, and other pests to a retail food establishment. Proper facilities and receptacles must be provided inside and outside the establishment to hold refuse, recyclables, and returnables that may accumulate. Refuse and garbage should be removed from retail food establishments frequently enough to minimize the development of objectionable odors and other conditions that attract or harbor insects, rodents, and other pests.

Discussion Questions (Short Answer)

1. What are some reasons customers use when selecting a retail food establishment to shop for food?
2. How can the appearance of a retail food establishment make a positive impression on customers?
3. Identify four characteristics that floors, walls, and ceilings should have when they are used in kitchens and warewashing areas.

4. List the primary components of a handwashing station.
5. How does the location of handwashing facilities influence sanitation and personal hygiene?
6. How does proper refuse and garbage disposal contribute to the pest control activities of a retail food establishment?

Quiz 7 (Multiple Choice)

1. Which of the following statements about toilet facilities is false?
   a. Toilet facilities must be available for all employees.
   b. Employee toilet facilities must be conveniently located and accessible to employees during all hours of operation.
   c. Separate toilet facilities should be provided for men and women.
   d. Poor sanitation in toilet facilities will influence customers' opinions about cleanliness but will not promote the spread of disease.
2. The most effective device for protecting the potable water system from contamination by backflow is a (an):
   a. Air gap.
   b. Double check valve.
   c. Hose bib.
   d. Vacuum breaker.
3. When an air gap is used, the vertical distance between the supply pipe (faucet) and the flood rim must be at least:
   a. Two times the diameter of the supply pipe, but never less than ½ inch.
   b. Two times the diameter of the supply pipe, but never less than 1 inch.
   c. Three times the diameter of the supply pipe, but never less than 2 inches.
   d. Four times the diameter of the supply pipe, but never less than 3 inches.

4. Which of the following statements about garbage and refuse sanitation is false?
   a. Proper disposal and storage of garbage and refuse are necessary to prevent contamination of food and equipment and avoid attracting insects and rodents.
   b. A trash receptacle should be provided in each area of the retail food establishment where refuse is generated.
   c. The equipment and receptacles used to store refuse, garbage, etc., must be durable, clean, nonabsorbcnt, lcak-proof, and pest-proof.
   d. Trash may be stored outdoors in plastic bags provided the bags are stored at least 15 inches off the ground.
5. You are surveying your establishment for unsafe conditions. Which one of the following situations requires corrective action?
   a. A trash receptacle with the lid off while in use.
   b. A handwashing station with a multi-use cloth towel for hand drying.
   c. Light colored ceramic tile being used for the walls of the food-preparation area.
   d. Anti-slip flooring provided in the warewashing area.
6. Backsiphonage is likely to occur if:
   a. The pressure in the potable water system drops below that of a nonpotable or contaminated water source.
   b. Contamination is forced into a potable water system through a connection that has a higher pressure than the water system.
   c. Pressure builds up in a sewer line due to blockage.
   d. The water seal in a kitchen trap is siphoned out.
7. The primary responsibility of retail food establishment managers in pest control is to ensure:
   a. Good sanitation that will eliminate food, water, and harborage areas.
   b. Pesticides are applied safely.
   c. The pest control operator they use employs integrated pest management.
   d. The parking area is kept free of litter.

8. The best way to encourage employees to wash their hands when needed is to:

    a. Provide a separate restroom for employees and customers.
    b. Provide properly equipped handwashing stations convenient to work areas.
    c. Provide hand sanitizers instead of handwashing lavatories in food preparation areas.
    d. Put up a sign in the employee locker room reminding them of the importance of proper handwashing.

9. Coving is a (an):

    a. Curved sealed edge between the floor and wall that eliminates sharp corners to make cleaning easier.
    b. Anti-slip floor covering used to protect employees from slips and falls.
    c. Plastic material used to seal cracks and crevices under and around equipment in a food establishment.
    d. Device used to prevent backsiphonage.

10. What factor has the greatest influence on where people choose to eat or shop for food?

    a. Cost of the food.
    b. Nutrition of food.
    c. Quality of service.
    d. Cleanliness of food and facilities.

Answers to the multiple choice questions are provided in Appendix A.

Suggested Reading/References

Bennett, G, W., J. W. Owens, and R.M. Corrigan (1997) *Truman's Scientific Guide to Pest Control Operations* (5th. Edition). Advanstar Communications. Cleveland, OH.

Food and Drug Administration and Conference for Food Protection (1997). *Food Establishment Plan Review Guide.* Washington, DC.

Kopanic, R. J., B.W. Sheldon, and C.G. Wright (1994). "Cockroaches as Vectors of *Salmonella*: Laboratory and Field Trials," *Journal of Food Protection,* Vol. 57., No. 2, pp. 125-32. February, 1994.

Longreè, K. and G. Armbruster (1996). *Quantity Food Sanitation.* Wiley. New York, NY.

Olsen, Alan R. (1998). "Regulatory Action Criteria for Filth and Other Extraneous Materials–Review of Flies and Foodborne Enteric Diseases," *Regulatory Toxicology and Pharmacology*, 28: 199-211.

Suggested Web Sites

Actron, Inc.
www.actroninc.com

American Society of Sanitary Engineering
www.asse-plumbing.org

Association of Applied Insect Ecologists
www.aaie.com

B&G Equipment
www.bgequip.com

Ecolab
www.ecolab.com

EPA Office of Prevention, Pesticides and Toxic Substances
www.epa.gov/internet/oppts

The Food Marketing Institute
www.fmi.org

Gateway to Government Food Safety Information
www.foodsafety.gov

Hobart Corporation
www.hobartcorp.com

Insect-O-Cutor
www.insect-o-cutor.com

Killgerm/PestWest, USA
www.Pestwest.com

National Pest Management Association
www.pestworld.org

The Orkin Company
www.orkin.com

PCO
www.pco.ca

Pest Control Industry
www.pestweb.com

Plumbnet-Industry Resources
www.plumbnet.com/resources.html

Steritech
www.stertech.com

Do-It Yourself Pest Control
www.doyourownpestcontrol.com

Virginia Polytechnic Institue and University Pesticide Programs
www.vtpp.ext.vt.edu

NSF International
www.nsf.org

Underwriters Labratories, Inc.
www.ul.com

United States Department of Agriculture (USDA)
www.usda.gov

United States Food and Drug Administration (FDA)
www.fda.gov

8 Accident Prevention and Crisis Management

Tornado Damage Causes Problems for Retail Food Establishment

A tornado touched down in a small Midwestern town and completely disrupted electric power and drinking water supplies. Electric lines were down throughout the town and water pipes were ruptured, pouring thousands of gallons of water into the local streets. The retail food establishments in the town were immediately affected and had to operate without these key utility services.

Luckily, the manager of the most popular retail food establishment was prepared for emergencies and had a crisis management plan immediately put into action. State and local health authorities were contacted for additional help. The plan included use of bottled water and single-service utensils. What else would the manager want to do?

Learning Objectives

After reading this chapter, you should be able to:

- ▲ Identify procedures for managing accidents.
- ▲ Discuss the importance of a safety audit.
- ▲ Recognize the value of safety in-service training.
- ▲ Identify public health rules and regulations that pertain to accidents and crisis management.
- ▲ Review a first-aid plan.
- ▲ Discuss the need for a fire exit plan, fire drill practice, and how to use a fire extinguisher.
- ▲ Inform staff of all terms used to alert for emergencies, i.e., "Code Adam."

- Identify the types and required locations of fire extinguishers.
- Post a list of emergency phone numbers by the telephone.
- Identify the responsibility to comply with Occupational Safety and Health Administration (OSHA) rules.
- Review procedures for crisis management due to loss of utilities.

Essential Term

Accidents

Cardiopulmonary resuscitation (CPR)

Safety in Retail Food Establishments

Retail food establishments contain many potential hazards that can cause **accidents** and injuries. Falls, cuts, scrapes, puncture wounds, and burns occur when food employees become careless or equipment is not in good repair. The main point to remember is accidents, like foodborne illnesses, can be prevented.

Accident prevention programs typically involve three types of activities:

- Eliminating hazards in the environment
- Providing personal protective equipment for workers
- Training employees about potential hazards and how to prevent or avoid them.

Eliminate hazards whenever possible. Use safety devices on equipment and keep all equipment in proper working order. A clean area free of clutter and possible hazards protects employees and customers.

Built-in safety guards keep hands away from moving blades during grinding and slicing operations. Prevent cuts when using slicers and deboning equipment by wearing steel and nylon woven fiber gloves. Hot pads protect hands against burns. Work shoes with closed toes and rubber soles shield the employee's feet from injury.

Employees must be taught how to prevent and avoid potential hazards that cause injury. They must know such things as:

- When and how to use ladders for climbing
- How to lift things properly

- ▲ How to use large heavy machinery such as a forklift, floor cleaning machine, compactor, etc.
- ▲ Why safety devices must be left in place on equipment
- ▲ How to dress for safety
- ▲ Minors are restricted from operating most power and mechanical equipment.

Before they start to work, give employees instructions on specific procedures and safety measures related to the assigned tasks. An ounce of prevention is worth far more than a pound of cure.

Each department needs a basic first-aid kit (Figure 8.1) placed in a readily accessible area but away from all food, equipment, utensils, linens, and single-use articles. The kit should include an instruction manual describing basic first-aid treatments, sterile dressings, adhesive tape and bandages, burn cream, antiseptic ointment, and other basic first-aid supplies. Check expiration dates on supplies monthly and replace them as needed. Post a list of important telephone numbers such as fire department, police, emergency medical system, and nearest hospital emergency room beside the telephone in a readily accessible area.

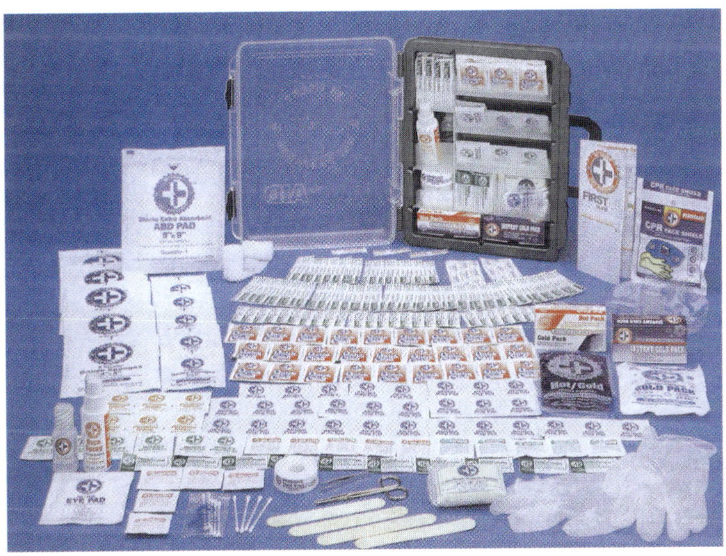

Figure 8.1 First-Aid Supplies

What to Do if an Accident Occurs

1. Stay calm. Information is needed to determine the seriousness of the injury. Begin by checking to see if the victim is responsive. Kneel and ask, "Are you OK? Would you like me to call for emergency medical services?"
2. Examine the injury and decide whether outside help is needed. If in doubt, call for help. Victims who are not breathing and do not have a heartbeat have a much greater chance of survival if they receive prompt medical care in a hospital or by trained paramedics.
3. Administer first aid according to the type of injury and your level of first-aid training.
4. Keep unnecessary personnel away from the victim.
5. Record the victim's name, the date and time of accident, type of injury or illness, any treatment given, and amount of time it takes for emergency assistance to arrive.

Common Types of Injuries

Falls

Falls are preventable. Keep floors clean and free from debris and spills. Wet floors, grease, or spilled food are hazards. Use signs to alert others of newly mopped wet floors (Figure 8.2). Use anti-slip mats to cover areas that have high traffic and are frequently wet. If someone falls, ask him or her to stay still, then determine the extent of injury. Seek medical assistance if there is a question about the victim's ability to move. Improper use of ladders or climbing on chairs, boxes, etc., instead of using a ladder can cause employee falls. Always use the proper equipment according to directions and intended use.

Injuries can be caused when items such as cans, bottles, or boxes fall from shelves or piles of stacked items and strike employees or customers. When this occurs act promptly to assist the injured person and call for help. Keep the correct stock level and load to prevent this problem.

Figure 8.2 Proper Signage Can Prevent Injuries

Cuts

Cuts are common injuries in retail food establishments. Carelessness with box cutters, openers, knives, slicers, choppers, grinders, mixers, and broken glass cause most problems. Keep knives sharp and store separately from other utensils. Always wear an approved cut resistant glove when working with knives or cutting equipment (Figure 8.3). Accidents occur frequently when dull tools slip and cause an injury. Never leave knives on counters or submerged in a sink full of soapy water. Always use the proper tool for the task. Do not try to catch a falling knife. Let it fall to the floor. Do not try to clean the circular blade of a slicer while the equipment is still assembled. *Always make sure slicers and other cutting equipment are unplugged before taking them apart for cleaning.*

Protect yourself from contact with body fluid from others when treating a cut. First find out the extent of the wound. Then apply pressure to the site with a clean cloth or towel and elevate the injury. If the cut is minor, wash thoroughly with mild soap and water. If available, apply antibiotic ointment. Apply a clean, water-resistant bandage. If the person will return to food preparation duties, have them wear disposable gloves to provide an extra layer of protection. More severe cuts should be checked by a medical professional. Report any injury to your loss prevention manager or immediate supervisor.

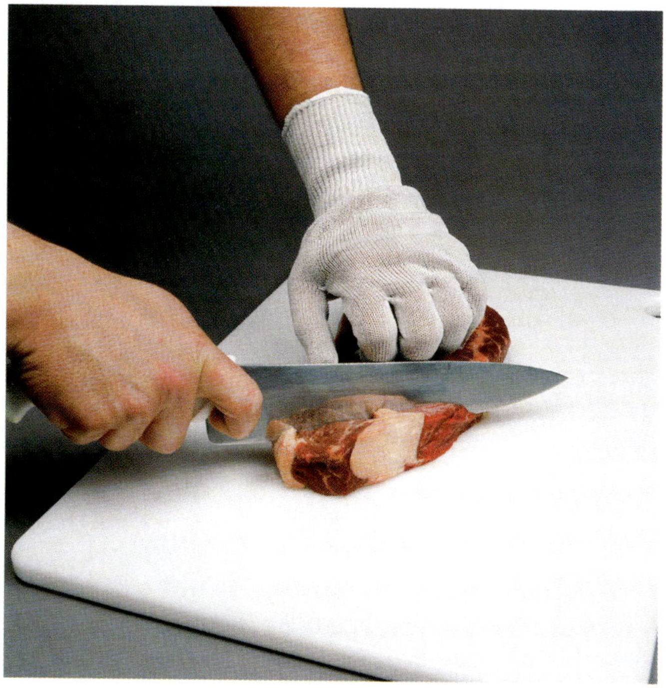

Figure 8.3 Protective Gloves Can Prevent Hand Injuries

Any food contaminated by blood or body fluids must be disposed of properly. Wash, rinse, and sanitize equipment, utensils, and work area where an injury occurred.

Always use power equipment according to the manufacturer's directions. Leave guard devices in place–they protect you from injury. Most current equipment, such as slicers and grinders, have tension or safety switches. This means the equipment runs when the switch is manually held in place. Do not apply tape or any other type of bypass. The system is there to protect the user.

A dead man's switch may be found in some large pieces of equipment, such as mixers, grinders, and cutters. A dead man's switch allows equipment to operate only when it is engaged. If the switch is released, the equipment will shut off immediately, thereby reducing the risk of injury. Usually the switches are located so you have to stretch your arms to each side to maintain the pressure. Thus, you cannot accidentally grab for something in or on the machine without releasing the switch and the machine stops. Overriding the switch with duct tape or other such material is illegal. Caution employees against such actions. The switch is there to protect them from serious harm.

Burns

Burns are classified as first degree (redness and pain), second degree (blisters, redness, and pain), or third degree (charring of skin layers, little or no pain). The extent of a burn is important. If a small area is burned, either first or second degree, first aid is usually all that is needed. Third-degree burns must be checked by a medical professional. Burn treatment includes the following:

- Remove the source of the burn. If the burn is caused by an electrical source, rescuers must first determine if the electricity is off before touching the victim.
- Soak the burned area in cool water to soothe minor burns and lower skin temperature.
- Over-the-counter pain medications may be used to help relieve pain and reduce inflammation and swelling.
- Cover the burn with a sterile bandage or a clean dry dressing. Do not puncture blisters.
- Keep the victim calm and quiet.

If a victim receives a second- or third-degree burn or if the burn covers a large area, you should seek medical assistance immediately. Burns on the face, hands, or groin area are considered serious and should be checked by a medical professional.

Chemical burns require immediate treatment. Call for medical help and check the Material Safety Data Sheet (MSDS) for emergency interventions. Chemical burns are often caused by mishandling dishwashing compounds, cleaning solvents, scrubbing machines, and batteries.

Poisoning

Poisoning can occur when food contaminated by chemical substances is inhaled, consumed, or sold. *If there is any question a food is contaminated by chemicals, do not wait.* Call for medical assistance immediately and begin internal recall crisis management. Contact the local health department if products have left premises. Call Poison Control Centers and follow instructions until help arrives.

Back Injuries

A class in how to properly lift objects can save a retail food establishment thousands of dollars. Back injuries cost the employer in lost productivity, medical care, and worker's compensation. A simple class and demonstration of good lifting and bending techniques is always a good investment.

The Food Marketing Institute has a series of videos on basic safety that is clear and easy to understand. The set includes a video on lifting and bending properly. Insurance companies and local health departments have resources to help educate employees on how to protect themselves from harm. Emergency medical service departments are often willing to come to the establishment to teach employees how to prevent injury.

Blood and Body Fluid Borne Diseases

Any type of human blood must be handled with care. AIDS (Acquired Immunodeficiency Syndrome) and Hepatitis B are both blood-borne pathogens that can infect people. The microbe enters the blood stream though contact with mucous membranes or open wounds in the skin.

Body fluids and feces also carry harmful microbes and must be disposed of carefully. There are special procedures called Universal Precautions to protect employees from contact with potentially harmful microbes. The Food Marketing Institute materials or local health care facilities can supply instruction in special techniques required to protect humans.

Employees must wear protective gloves and outer garments and use cleaning and sanitizing procedures to remove spills. A solution of 1 part bleach to 10 parts water or other commercial sanitizer approved for this purpose should be used to disinfect any area contaminated with blood. Do not attempt to clean up a spill or assist another person without first obtaining gloves and protective gear for yourself.

Broken glass or other sharp objects should not be handled, even with gloved hands. Use a broom, brush, and dust pan to remove them to a designated container for special handling and disposal. When gloves are removed, pull the glove off turning it inside out. Drop it in the container with the contaminated material and then thoroughly scrub hands and forearms under warm running water with soap. Any clothing that has been contaminated must be removed and washed. Check for any other skin contact and immediately flush and wash away splatters.

Federal law (OSHA 29CFR1910.i030) mandates employees should see a medical professional for care and follow up. Always create an incident report to describe time, date, type of contact, and actions taken to protect the employee. The Food Marketing Institute has produced a video and learning materials about blood-borne pathogens and how to manage cleanup, disposal, and aftercare.

Employee Medications

According to the *FDA Food Code*, only those medications necessary for the employee's health are allowed in a retail food establishment. This does not apply to medicines stored or displayed for retail sale. Employee medication must be clearly labeled and stored in an area away from food, equipment, utensils, linens, and single-use items like straws, eating utensils, and napkins. Medications stored in a refrigerator where food is stored must be kept in a clearly labeled and covered container located on the lowest shelf of the refrigerator. Medications cannot be stored in display or storage cases, or walk-in units.

Cardiopulmonary Resuscitation (CPR) and First Aid for Choking

It is recommended that your staff know how to administer first aid for choking or **Cardiopulmonary Resuscitation (CPR)**. However, posting charts with instructions for these procedures and other lifesaving steps can be helpful in emergencies (Figure 8.4). These charts are available from the American Red Cross and the American Heart Association. It is desirable to have at least one person on each shift trained in emergency response procedures and first aid. Remember, knowing how to do the right thing at the right moment can save a life.

Local public health departments, American Red Cross, American Heart Association, National Safety Council, insurance companies, emergency medical systems, fire departments, and hospitals can provide classes, demonstrations, posters, videos, and other materials for use in retail food establishments. Personnel who volunteer to take safety related courses for use in the establishment are great assets to the establishment.

Self-Inspection Safety Checks

Just as self-inspection programs help identify food safety problems in retail food establishments, safety audits can help prevent accidents. Conducting a safety audit can be one of the most cost effective tasks you can perform. Fire and accident prevention can result in considerable cost savings.

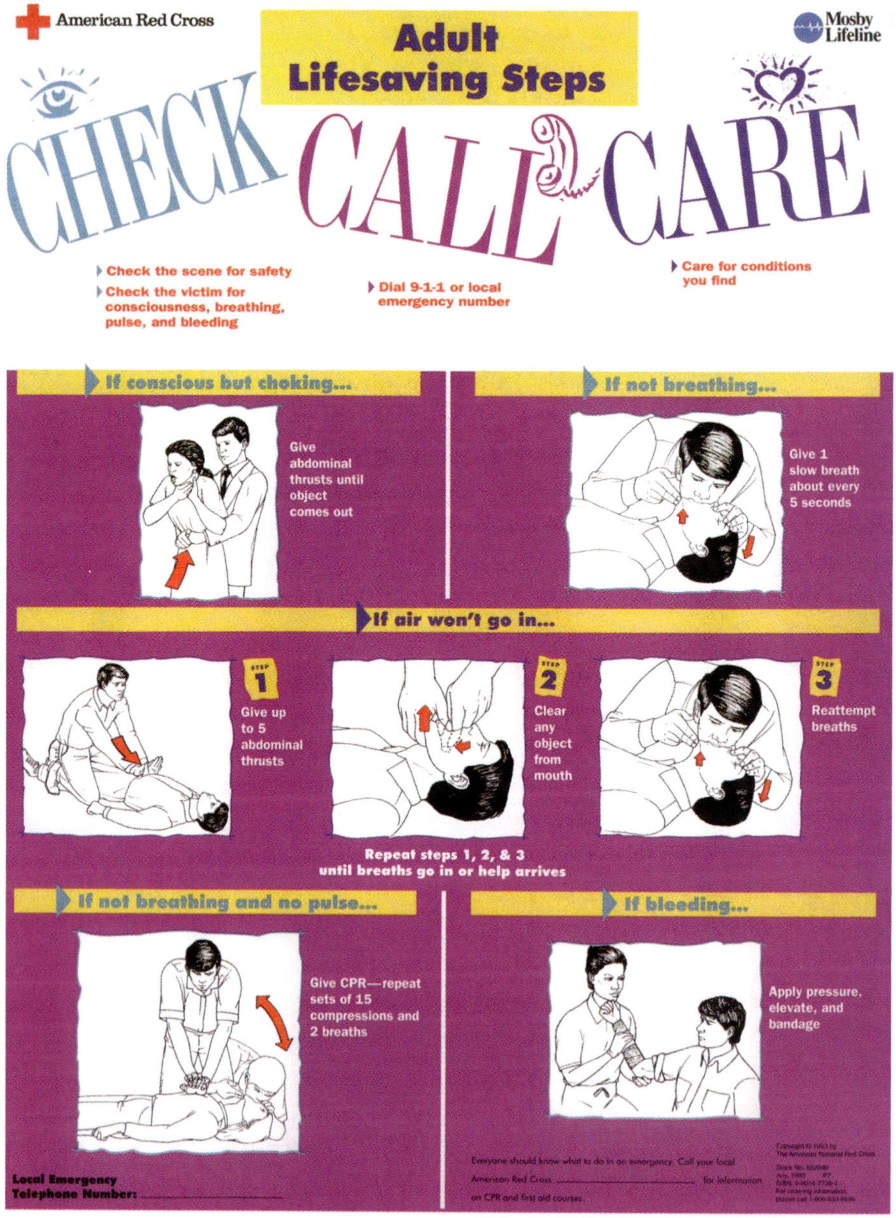

Figure 8.4 Lifesaving Steps
(Source: American Red Cross - used with permission)

Incident Reports

When an accident does occur, it is always advisable to fill out an incident report. Insurance companies require this documentation when processing claims. Record the time of the incident, conditions that contributed to the problem, extent of observed injury, any treatment provided at the establishment or a medical facility, and prescribed aftercare. Notify insurance companies as soon as possible after an accident has occurred.

Fire Safety

Grease and electrical problems are common causes of fire in retail food establishments. Regular inspections of electrical equipment and prompt repair of faulty wiring or equipment can decrease this problem. Avoid overloading circuits and improper venting or grounding. Teach and monitor employees on the hazards of pulling electrical plugs out of sockets by the cord. They must grasp the plug head and then pull.

Keep hoods, ventilating fans, filters, and adjoining walls clean and clear of greasy buildup. Perform regular maintenance according to the manufacturer's directions.

In case of a fire emergency, the first thing to do is get customers and employees out of the building. Be sure you evacuate the building and call the fire department.

Fire Extinguishers

Portable fire extinguishers must be operable, easy to find, and employed according to type of fire (Figure 8.5).

Type of Fire Extinguisher	Material Burning	Most Common Type of Extinguisher Used
Type A	wood, paper, cloth	Pressurized water or ABC
Type B	grease, gasoline, solvents	COx or B or BC
Type C	motors, switches, electrical	COx or C

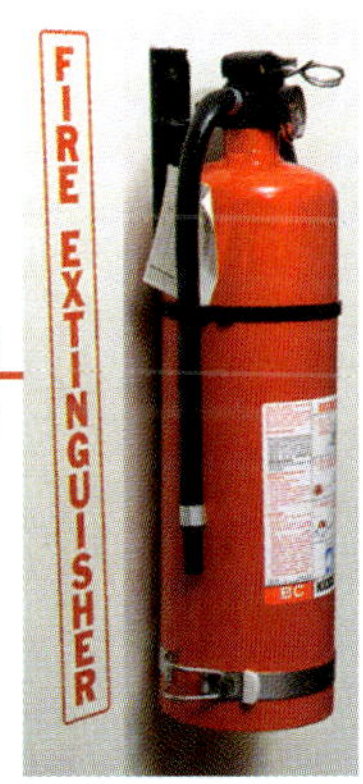

Figure 8.5 Types of Fire Extinguishers

Type A or pressurized water is not recommended for food production areas because it might be used on a grease fire by mistake. Grease floats on water and would easily cause the fire to spread. Water on electrical fires could cause potential shock to personnel and extensive damage, too. Carbon dioxide (CO_2) or type B/C extinguishers are recommended because it is safe and leaves no damaging residue on machinery or food-contact surfaces. Any exposed food must always be discarded. ABC multi-purpose extinguishers do leave a residue. Local fire departments can provide information and demonstrations on the proper type of fire safety equipment and how to use it.

Because employees' clothing can ignite when working around stoves or a heat source, a safety blanket should be stored in an area where it can easily be reached if needed. Use the blanket to smother the flames and then call for help.

If you detect a fire you cannot extinguish quickly, call for help. Then shut off exhaust fans, turn off equipment, close all doors and windows, and evacuate the area. Every retail food establishment needs a fire safety plan. Clearly mark routes of evacuation. Hold fire drills at periodic intervals to ensure all employees know how and where to exit if the building is on fire. Battery powered emergency lights are required in most jurisdictions. They automatically provide light when power is interrupted.

Fire safety inspectors check fire extinguishers to certify them as operable. Portable fire extinguishers should be checked or serviced at least annually and marked with tags that specify the next due date for service. Fines can be levied for out-of-date equipment. If an extinguisher weighs less than 40 pounds, it can be mounted 5 feet from the ground. If it weighs more than 40 pounds, it may be stored no more than 3½ feet off the ground.

Hood Systems

Hood systems are usually installed over cooking equipment and dishwashers to remove vapors, heat, and smoke. Some hood canopies may have a built-in extinguishing agent. These devices can effectively extinguish a fire. Employees need to know how they work. Hood systems should be installed and periodically serviced by fire prevention professionals, typically twice a year.

Sprinkler Systems

Sprinkler systems are often required by law in new or remodeled buildings. Supplies must be stored the regulated distance from sprinkler heads. If a sprinkler head is accidentally broken off, contact your sprinkler service immediately.

Operating a Forklift

Forklifts are common pieces of equipment in large retail food establishments. Any employee who operates a forklift must be certified through the Occupational Health and Safety Administration. Just having a driver's license is not enough to use this equipment. Because this piece of machinery can easily become unbalanced and topple over or injure someone who may be standing under the lift forks, operators need to demonstrate knowledge of special handling techniques.

Always check operating manuals and local rules before assigning personnel. If rules are violated, problems can easily become difficult to manage.

Crisis Management

A crisis can occur at any time. Besides natural disasters such as fire, flood, storms, or earthquakes, retail food establishments must deal with interference in normal operations that can literally cripple production. Other types of crises that can have a significant impact on retail food operations are power outages, interruption of water and sewer service, an outbreak of foodborne illness, media investigations and unexpected loss of personnel. How do you handle difficult situations? Develop a plan to handle the crisis with a problem-solving approach.

First, evaluate the extent of the problem. Check the facts and decide what possible outcomes can be expected. Are you going to need outside help? If you are working with a group of employees and supervisors, analyze their talents and use a team approach. One spokesperson is better than having too many individuals giving out information. Maintain control within your organization.

Most retail food establishments have a system of codes to alert employees about certain conditions. One increasingly and sometimes mandated by state or local law is "Code Adam." This refers to a lock-down procedure to keep everyone in the building. If a child is reported missing, this protects the child from abduction or accidentally leaving the building without a caretaker (parent, guardian, etc.).

Set priorities according to the resources that exist. Remain calm. Look for positives. Follow established rules and regulations as set up in codes. Identify any outside resources that can help solve the problem. Keep a record of actions and communications in case it should be needed. In all situations, be honest.

Water Supply Emergency Procedures

On occasion a water utility will issue a boil water order when the bacteriological quality of the water it provides does not meet the requirements of the Safe Drinking Water Act. This commonly happens when a water main breaks or when the source of the drinking water is contaminated by runoff and other sources of pollution.

To continue operating under "boil water advisories/notices" or "interrupted water service" from all water supplies, retail food establishments must secure and use safe drinking water from an approved source. This can be bottled water or water hauled to the site in tank trucks. Bottled drinking water used or sold in a retail food establishment must be obtained from approved sources in accordance with the Code of Federal Regulations (21CFR 129 - Processing and Bottling of Bottled Drinking Water).

In emergencies, or as a temporary measure, water from contaminated or suspect sources can be disinfected by either chlorination or boiling. To chlorinate water, add six drops of liquid chlorine bleach to one gallon of water and mix. Check the label on the container to make sure the active ingredient, sodium hypochlorite, is available at 5.25 ppm. The important things to remember when chlorinating water are:

- ▲ Wait thirty (30) minutes after adding the chlorine before using the water for drinking or cooking purposes
- ▲ If the six drops of bleach does not give the water a slight taste of chlorine, double the amount of chlorine until a slight taste is present. The taste of chlorine will be evidence the water is safe to drink.

Water may also be purified by boiling. Bring the water to a full boil for at least five minutes. Cool and aerate the boiled water by pouring it from one clean container to another. This will reduce the flat taste caused by boiling.

Retail food establishments may consider the following alternative procedures to minimize water usage:

- Commercially packaged ice may be substituted for ice made on site
- Single-service items or disposable utensils may be substituted for reusable dishes and utensils
- Prepared foods from approved sources in place of complex preparation on site
- Restrict menu choices or hours of operation
- Portable toilets may be utilized for sanitary purposes.

After the water emergency is officially lifted or water service resumes, these precautionary measures must be followed:

- Flush the building water lines and clean faucet screens and water line strainers on mechanical dishwashing machines and similar equipment
- Flush and sanitize all water-using fixtures and appliances such as ice machines, beverage dispensers, hot water heaters, etc.
- Clean and sanitize all fixtures, sinks, and equipment connected to water lines.

There must be water pressure before resuming operations in a retail food establishment, and the water should be sampled for bacteriological quality. The safety of water cannot be judged by color, odor, or taste. Don't hesitate to contact your local regulatory agency if you have any questions regarding appropriate operations at your establishment during water emergency orders.

Bioterrorism

On September 11, 2001, terrorism in the United States became a reality. For the past several years, the federal government and state agencies have developed plans to react to potential efforts of sabotage. Reacting in a rational manner is extremely important. The Food Marketing Institute (FMI) is developing guidelines to handle any suspected tampering or contamination problem in retail food establishments. There are several actions that can be used to manage these types of problems. These include the following:

- Monitor employees and allow only approved employees into production areas. The use of photo identification name tags helps to identify employees.

- ▲ Protect food preparation areas from anyone who is not assigned to that area.
- ▲ Review visitor policies and change rules to keep areas secure. Use sign-in and sign-out logs.
- ▲ Prohibit personal items like lunch containers, cases, purses, and other such items from processing areas. Provide storage for employee property in a separate room or locker.
- ▲ Report any unusual or suspicious activity to your supervisor or manager.
- ▲ If a suspicious problem is identified, call the FBI and FDA Office of Crime Investigation. These agencies are best equipped to handle emergency situations and will advise you of the steps to follow.
- ▲ Designate a spokesperson to deal with media or other inquiries. This method ensures information is managed carefully and mixed messages are avoided.

Product tampering is also a food safety concern, and many retail food establishments have developed policies and procedures to deal with such problems. Managers should teach employees to be alert and to report any unusual activity immediately. It is always better to be overly cautious than to ignore a potential incident.

FMI has a *Pre-Emergency Planning and Disaster Recovery Manual* with phone numbers and instructions on what to do in cases of emergency. For more information about this manual, you may call the FMI Publications and Video Sales Department at (202) 220-0723.

FMI can also provide assistance and information for use when implementing product recalls. For help on recalls, call the Food Safety Department at (202) 223-0661. FMI also has a document called "Measures to Improve Security in the Supermarket Industry." For more information on store security, call the above number.

The Centers for Disease Control and Prevention has prepared large amounts of information on bioterrorism and proper protocols for handling suspected biological threats and communicable diseases. Log on to the CDC Website at www.cdc.gov for information about how to manage bioterrorism and who to contact if threats are suspected. FDA has a Web site at www.fda.gov for bioterrorism assistance.

Foodborne Illness Incident or Outbreak

If dealing with a suspected foodborne illness problem, cooperate with the customers and seek help from your local food regulatory agency. Work with the local regulatory agency to determine what may have caused the customer's illness. Establish the time the suspect food was purchased and eaten. If a sample of the product is still available, preserve it for laboratory analysis. Ask the customer if medical attention has been sought and to save samples of any product that remain. Initiate your crisis management team members, which should include team members from your media affairs, food safety, loss prevention, and legal counsel. Remove all affected food from the sales areas, isolate it in a separate area, and label it so it does not become mixed with other foods. Do not sell any more of the suspect product. Listen to everything the customer says and look for clues that might explain the problem. Do not try to diagnose the problem or belittle the customer's complaint. Keep a record of the interview. Assure the customer you will get back to him or her after investigating the problem. Make sure you do contact the customer again after facts are gathered. Do not admit your establishment is at fault–you have no proof of that until facts are checked.

Check the Product Source

If the reported item was prepared in the store, remove it from sale immediately. If two or more persons report the same problem, consult outside resources (local health department) on how to proceed. When preparing food samples for testing, it is wise to keep a sample in your freezer for cross validation of laboratory results. Public health officials, insurance agents, and, if needed, an attorney can help you manage the problem.

Formal Investigations

Remove suspect items from use. Health inspectors may want to conduct an inspection at the establishment. Employees may be required to have medical examinations. Closing a facility for a period of time for cleaning may be needed. Keep in mind whatever must be done should be accomplished as quickly as possible. Cooperation to protect public health is the end goal. In the final analysis, it does not matter where the error occurred. It must be corrected and prevented from happening again.

Whatever the crisis may be, use a calm, systematic approach and evaluate actions as much as possible. Crisis management is not an easy task. It is always useful to have some kind of protocol to guide you through the event. Put together a team or committee consisting of management and employees to regularly brainstorm worst case scenarios. Plan disaster drills to keep employees informed and trained in crisis management skills. Designate a spokesperson who is responsible for talking to the media, should they become involved. Keep a notebook with instructions on how to proceed in emergency situations.

Summary

Retail food establishments must be prepared to deal with unexpected events such as accidents, severe weather, and interruption of utility service. In the case presented at the beginning of the chapter, the food establishment is faced with a disruption of water supply to the establishment. This might have occurred due to a water main break, a fire in the vicinity, or some other cause. When faced with a water outage, the managers and employees of a retail food establishment must determine how to best cope with the situation. Under extreme conditions, the retail food establishment may have to be closed until service is restored. In less dire circumstances, other measures may be required. Sales may be restricted to sales of prepackaged and prepared foods. Disposable plates, cups and utensils may be used. In all circumstances, customer safety must be the top priority of the retail food establishment. Never make a decision that will put the safety and well-being of your customers in jeopardy.

Accident prevention programs are good investments. Use community resources to help build an effective plan. Insurance companies and vendors can supply training aids on accident prevention. Fire departments and emergency healthcare providers can provide demonstrations and advice that prepares you for accidents or other problems linked to safety. If you never need the plan, consider yourself lucky. However, be prepared.

Check the emergency lights to see if they are on as required. All equipment with on/off switches should be turned off until power is restored. Take care of all perishable items according to procedures. Throwaway items must be disposed of according to health department rules to keep customers from obtaining discarded products that could cause harm. The Food Marketing Institute [*Pre-Emergency Planning and Disaster Recovery Manual (1996)*] contains detailed instruction to manage all refrigerated, frozen, and dairy product materials during loss of power.

Quiz 8 (Multiple Choice)

1. When slicing or removing bones from fresh meat, the food employee should:
   a. Always wear single-use plastic gloves.
   b. Use a fork to stabilize the product.
   c. Wear nylon or metal mesh gloves.
   d. Hold the meat firmly with a bare hand.
2. When an employee brings prescribed medication to work that must be refrigerated, it should be
   a. Put in a closed and labeled container in the employee lounge refrigerator.
   b. Placed in the back corner of a refrigerated display case.
   c. Wrapped in plastic and put in the ice machine.
   d. Sent home with the employee because it is not allowed in the area.
3. The Occupational Safety and Health Administration requires all retail food establishments maintain a file with information on how to treat chemical exposures. These documents are
   a. Material Sample Detail Sheets.
   b. Medical Symptom Data Signs.
   c. Material Safety Data Sheets.
   d. Medical Sample Design Contents.
4. The document that lists when an accident occurs and what action was taken is called an.
   a. Incident report.
   b. Evaluation record.
   c. Accident record.
   d. Insurance form.
5. Every retail food establishment must have the appropriate
   a. Number of shelves.
   b. Fire extinguishers.
   c. Inventory control system.
   d. Number of check-out stations.

Answers to the multiple choice questions are provided in Appendix A.

Reference/Suggested Readings

Food and Drug Administration (2001). *2001 Food Code.* United States Public Health Service. Washington, DC.

Food Marketing Institute (1996). *Pre-Emergency Planning and Disaster Recovery Manual.* Loss Prevention Services Department. Washington, DC.

McSwane, D., Rue, N., Linton, R. (2003). *Essentials of Food Safety and Sanitation.* Prentice Hall, Upper Saddle River, NJ.

Suggested Web Sites

American College of Emergency Physicians
www.acep.org

American Red Cross
www.redcross.org

Food Marketing Institute
www.fmi.org

Health Answers Medical Reference Library
www.healthanswers.com

Injury Control Resource Information Network
www.injurycontrol.com

Medicine Net Company
www.medicinenet.com

National Safety Council
www.nsc.org

Occupational Safety and Health Administration
www.osha.gov

Safety Online
www.safetyonline.net

Safety Smart Company
www.safetytalks.com

Vermont Safety Information Resources, Inc.
www.hazard.com

Gateway to Government Food Safety Information
www.foodsafety.gov

9 EDUCATION AND TRAINING

A Case of Foodborne Illness a Trainer Could Have Prevented

A new employee was eager to begin working at his job in a retail food establishment. There was just one problem. He did not speak English well enough to really understand what the trainer was explaining about his job.

The trainer was rushing to get a report finished when asked to help the employee learn how to set up the fruit display in the produce department. A packet of translated training materials had just arrived, and the trainer decided this was the perfect time to use them. She handed the booklet to the employee and asked if he understood what she wanted. Anxious to prove how quickly he could get to work, the employee nodded enthusiastically and accepted the pamphlet. When he opened the first page, his heart sank. It was all in words. He had hoped there would be enough pictures to help him understand what to do. He could not read, even in his own language.

Unable to let anyone know about his problem because he feared he might lose the job, he watched other clerks to imitate their actions. After an hour of stacking fruit, he needed to go to the toilet. His problem with diarrhea and abdominal pain had gotten worse since arriving in town.

After using the toilet, he rinsed his hands but did not know how to work the soap dispenser or electric hand dryer on the wall. He quickly dried his hands on his shirt and hurried back to work, packaging grapes and setting up the apple display.

In the next week, the manager received several complaints from customers who were having symptoms of foodborne illness. One customer had been hospitalized and tests determined she was suffering from shigellosis. The health inspector came to the establishment and decided to test all employees for Shigella spp. *The new employee tested positive and so did three others who had worked with him. How could this problem have been avoided?*

Learning Objectives

After reading this chapter, you should be able to:

- ▲ Recognize the critical need for education and training in retail food establishments.
- ▲ Determine learner needs for each department within retail food establishments.
- ▲ Locate resources available for food safety and sanitation training including media, government materials, and the Internet.
- ▲ Develop evaluation tools to measure training outcomes.

Essential Terms

Needs assessment

Standards

Transfer of learning

Visual aids

Market Need

The topic of food safety and sanitation rarely stirs the imagination and interest of anyone. Your challenge–get the information across to all levels of learners who need the knowledge to perform successfully.

Errors by employees in food safety procedures cost real dollars. Retail food establishments have learned some hard lessons in the last five years when problems with foodborne illness became front-page news. Emerging pathogens, an informed public, more highly susceptible individuals, and a global food supply create daily challenges to the food industry. In previous chapters, you learned about these problems. Now, you must learn what you can do to prepare your employees to prevent foodborne illness.

Issues of cultural, diversity, age, nationality, ability, and motivation can make this a difficult job. Economic stability may or may not affect the number of people who want to work in retail food establishments. The only thing you can count on is uncertainty and change.

Learners

As a trainer, you must recognize the age specific difference of your audience. To ignore this fact sets you up for failure. A large number of employees in retail food establishments are between 16 and 24 years old. However, an increasing number of older adults, aged 50 to 70, are also finding employment in these establishments.

You must consider the mixed generational workforce as a factor when you design your training programs (Figure 9.1). If you want to be effective, you must recognize the age specific and learning style differences in your audience.

Figure 9.1 Training

What are we talking about when we discuss a mixed generation work group? The chart in Figure 9.2 is extrapolated from **Generations at Work**, by Ron Zemke, Claire Raines, and Bob Filiczak. It provides a quick overview of the mixed generation phenomenon.

Training involves many diverse activities and individuals. You may teach a group of managers and supervisors to prepare them to take a food safety certification examination one week and train employees how to perform a particular task in their department in another week.

Your training plan should take into consideration the cultural and age diversity of your audience to assure a successful training program.

Born:	Veterans 1922-1942 52 million	Baby Boomers 1943-1960 73.2 million	GenXers 1960-1980 70.1 million	Nexters 1980-2000 6.7 million
Work Ethic	Dedicated	Driven	Balanced	Determined
View of Authority	Respectful	Love/Hate	Unimpressed	Polite
Training Type	Traditional Classroom	Lifelong learning Scannable material Read selectively	Computer-based CD ROM Interactive Read less	Read more Learn Interactive Use more technology

Generations at Work, AMACON 2000

Figure 9.2 Mixed Generational Workforce

Current Information and Technical Challenges

Corporate trainers or consultants who contract for work face the challenge of keeping their knowledge base current. Each site also has individual problems related to geographic location, corporate culture, employee turnover, and customer mix. As more businesses consolidate, the "Mom and Pop" operations are becoming obsolete. Giant retailers are now starting chains of super stores that sell everything, including retail food items. These market trends affect the methods and delivery systems of traditional training.

When trainers go to major meetings or read trade journals, excitement over information technology, CD ROM interactive materials, Internet online instruction, and Internet webcast training build to a roar. Everyone flocks around displays of the latest tools and teaching toys.

But hold on! How about a reality check–consider the average learner and available resources at home base before rocketing off into the stratosphere. Every new methodology deserves review and consideration. Then think critically of what employees actually need and have available. Flexibility and creative solutions are needed.

Training kits, canned presentations, technology, and outside experts may help. However, if you are the trainer and must solve multiple problems, recognize the need to be well-informed and understanding when a plan for training is devised.

Training is Important

Training is expensive but consider the alternative. Repairing a problem costs time, money, and may cause serious harm to reputation and sales. The positive result of a well-trained staff cannot be ignored.

Higher retention of employees means fewer dollars spent in orientation and on-the-job training. Satisfied customers who trust the retail food establishment and its staff return again and again. Employee and customer accidents can also be prevented with proper training. This can keep down the cost of doing business. The list of benefits for proper training far outweighs the few dollars saved when training is ignored. Safe food handling will prevent foodborne illness and proper training is the only way to ensure that process.(Figure 9.3).

Figure 9.3 Training is Important

Barriers to Training

Why do so many retail food establishments resist educating employees? The benefits are obvious–what is the problem?

▲ **Perceived Lack of Time**

When an operation is busy, it seems there is too much to do to stop for class time. Any time off task is seen as a loss to productivity. Designated time needs to be set aside for training.

▲ **Lack of Commitment**

A manager that does not value training may not understand the benefits training provides for his or her operation. Time and money spent for training may seem like a negative expenditure, when it is actually an investment in future efficiency. The commitment for food safety training should be supported from top management and the importance of training should be stressed. This helps training be effective for all types of employees.

▲ **Training Skill Deficit**

Recognition of the lack of knowledge requires insight into the issues that need fixing. It is sometimes easier to blame the employee when they do something wrong than to figure out why the error occurred. An organization needs to assess the training skills and levels and constantly make an effort to improve training and skill performance and retention.

▲ **The Culture of the Organization**

If the organization does not recognize the benefits of training, it is difficult to change core values. Top management must commit resources and support before training can become a central benefit to the overall operation. Government intervention with mandated certification requirements might seem like an intrusion into private industry. However, forced change may be the only answer to some problems.

The Agenda for Training in Retail Food Establishments

Food safety and sanitation training programs in retail food establishments usually start with a needs assessment. This will include:

- ▲ Department personnel who will attend and the area in which they manage, supervise, or work such as the
  - ▼ Meat Department
  - ▼ Delicatessen Department
  - ▼ Produce Department
  - ▼ Bakery Department
  - ▼ General store merchandising
  - ▼ Receiving and Storage.
- ▲ How many learners will be in each group?
- ▲ What amount of time is allotted for the program?
- ▲ Where will the training take place?
- ▲ What teaching and learning resources are available?
- ▲ What is the average level of knowledge, skills, and abilities?
- ▲ Are there any language differences to be addressed?
- ▲ What outcomes are expected by upper-level management?
- ▲ Are there any current issues that need immediate attention?
- ▲ Do evaluation tools exist for this site or will you need to provide them?

Once you have answered these questions, a plan can be developed to conduct training. This approach is aimed at group presentations, usually scheduled for a half-, whole-, or two-day sessions. The length of time spent on training depends on the topic and outcomes that are expected.

Departments have different needs and materials that require specific instruction on operation and maintenance. Safe food practices–keep it hot, keep it cold, keep it clean–cross all lines. Application varies according to the items within the area. It takes time and patience to teach others how to teach. The motto "each one teach one" is a mainstay in vocational learning.

Needs Assessments

In education, the simple truth lies in knowing what must be taught and in choosing the right tools and times to do it. Get assistance when you need it. There is no shame in admitting you need help. Look for the expert that can get the job done.

Question employees about what they already know and address real life issues. Adults respond best to education that is relevant to their need to know how to do their work successfully. There is a difference between need to know and nice to know information.

What you plan to teach and how you plan to get information across depends on several factors. A **needs assessment** that helps determine the what, who, when, how and why of a training session should be implemented.

Every retail food establishment should teach employees about:

- Emergency situations and response systems
- Issues related to health and hygiene
- Common causes of foodborne illness
- Food safety management procedures that include time and temperature controls
- Cleaning and sanitation procedures
- How to safely use and clean equipment and utensils
- Accident prevention measures
- Customer relations and work place etiquette
- Company policies and procedures
- Department specific procedures and protocols
- Reportable illnesses and awareness.

This list contains general suggestions. Each trainer must adapt their needs to available resources, time, type of learner, and any other requirements that need to be addressed.

Best Practices Recommended Times for Training

Short sessions–30 to 45 minutes at a time–are easier to handle. Most adults prefer to learn a task or fact and then immediately try it out. Watching the learner apply what you teach is always an effective method to determine how much someone has learned. Remember, repetition will help learners memorize concepts.

When longer classes are needed, break up the time with activities and interaction that allows participants to get up and move around. Keep your training inside the outlined time frames; do not start early and do not run long. When a break time is scheduled, honor that commitment. If a time is posted for instruction, do not cancel unless absolutely necessary. Send the message that training is important and follow through.

Standard Operating Procedures Federal, State, and Local Rules and Regulations

Every retail food establishment must comply with the rules and regulations of federal, state, and local agencies. Retail establishments should have a procedure for standard operating procedures (SOPs) and they should be taught and practiced. Your training plan will need to include each element employees are responsible for using in their daily work.

Hazard Analysis Critical Control Point System

In the next chapter, you will learn about food safety management programs. Employees will need special instruction in how to use the HACCP food safety system in your establishment. Additional instruction may be needed to assure proper monitoring of critical limits and recording of information to verify the system is working properly.

Environment

Training may take place in a variety of settings. The topic, available space, number of students, and resources needed to accomplish the job must be considered. Lighting, ventilation, and the ability to see and hear what is being taught. The number of variables can be surprising.

Physical comfort of the learner is important. Whenever possible, arrange for maximum space and seating. Keep an eye on your learners. Are they getting sleepy? Turning down the light for a slide, video, or PowerPoint® presentation may seem the way to go, but watch out for the effect a darkened room can have on listeners.

Do you have their attention? Always "read" the faces of those you are teaching to see how well they are listening. Ask questions to get feedback. Stimulate discussions to find out what the learner knows or wants to share. Often, the people who do the job have some of the best solutions. Listen to what they have to say. Control your class, whether it is small, medium, or large.

Conduct the Training

When the training plan is complete, the next step is to conduct the training. An effective training session includes the following points:

▲ **Keep It Simple**

Do not try to cram too much information into one session. In the end, the learner is overwhelmed and frustrated. Focus on the main points you are trying to share with employees. Do not stray from your focus. For example, to teach someone proper handwashing techniques, stick to the steps and present them in order. Do not start talking about grooming and uniform standards as well. Save that for another short training module or discussion.

▲ **Make It Fun**

Training should be an enjoyable experience for both the trainer and trainee. When learning is fun, the participants remember more of what was taught. Use colored pictures and get the trainees involved as much as possible. Include a team activity or energizer in your training session. Simple games and small prizes can help make the class fun and will reinforce learning.

▲ **Involve the Audience**

Employees want to be involved in the learning. Ask open-ended questions, use small group discussions, play games, have quizzes, and include them in demonstrations. These are ways to get your participants involved. Use pre- and post-tests to evaluate the outcomes of the training.

▲ **Use Visual Aids**

We are a highly visual society. Television has taught us the value of pictures and diagrams to make a point. To enhance learning in food settings, it is important to use **visual aids** in training. Videos, charts, posters, models, slides, etc., are effective visual aids. A Training Kit has been developed to accompany this book. You may obtain a copy by calling FMI Publications and Video Sales Department at (202) 220-0723. Make sure any visual aids are large enough for your trainees to see. If you use a flip chart and write too small, it frustrates the trainee because they cannot see what is going on. Training is a combination of a skilled trainer and appropriate visual aids. Use visual aids as an enhancement, not a replacement for instruction.

Technology

See Figure 9.2 Mixed Generational Workforce on Page 324 on how generations respond to different technology. Individualized learning on a computer or a video program may seem the perfect answer, but can the learner handle the technique? Marvels of modern communication are wonderful but an instructor is still a necessary part of the learning process.

The other problem may be lack of tools to use advanced methodology. If you have no computer, video player, or other advanced equipment, older methods will be needed.

Evaluation

All orientation to the job and task training should focus on the changes in behavior required to perform work correctly. Even new employees who come from similar situations and report they know how to do the job will need evaluation. Each retail food unit has a particular method of operation. The **standards** for performance and expected outcomes supply the points for evaluation.

Keep a training record for all classes given and an individual card or file for each employee. While this may be a time-consuming effort, it always pays off. Individual training moments given one-on-one may not get recorded but a card in your pocket or a little notebook with a few notes may help prove a point on individual employee performance. One quick method to teach is when you or another employee shows an employee how to perform a task. Then, you ask the employee to do the same thing. Instant feedback proves if there was a transfer of knowledge.

Effective evaluation tools are derived from the task analysis of employee competencies. Organizations may use a standard format for performance with a rating scale for general behaviors. Add to that specific lists of job tasks and attitude toward team members or associates in the same department.

No one likes to do an annual evaluation. It is time-consuming and requires a record of actions from previous weeks or months. However, it is invaluable in decision making and can target star performers who deserve extra recognition or poor employees who need more training or motivation. Work environments that encourage and reward correct behavior will benefit from the effort.

Transfers, promotions, raises, disciplinary actions, and firing of incompetent employees need documentation for validity. Never underestimate the value of this process. It is often tedious and time-consuming, but necessary.

Training Moments

In addition to scheduled training sessions, remember learning can occur continuously in retail food establishments. Through manager and trainer actions, employees learn proper behavior. Be a positive role model for your employees. Use training moments to improve performance. When employees are not practicing proper procedures, be a coach. Teach them the right way to do it. Reinforce good habits. Reward employees who perform their jobs correctly. Look for learning readiness. Adults must see relevance in what they are asked to learn. Once they accept the reason for training, they usually remember it well.

Summary

The situation at the beginning of this chapter was designed to highlight the importance of never taking for granted that someone knows what to do in all situations. Never assume new employees are able to read–even in their native language. A good example of the use of pictures and diagrams is found on airlines. They no longer rely on written emergency instructions. The international use of pictures and symbols has become a necessary method of communication. Charts, pictures, assigning a new employee to a trained employee–all of these strategies are necessary in today's world.

Education and training cost time and money but always consider the consequences. A culturally diverse workforce and changing economic trends create new challenges every day. More basic education takes place in the work place today than it did 25 years ago. This trend will only accelerate in the years to come. Remember, the best defense is a good offense. Spend wisely.

Quiz 9 (Multiple Choice)

1. Today's food industry leaders understand training is:
   a. A cost without benefits.
   b. Not necessary.
   c. An investment.
   d. A one-time event.
2. One of the benefits of effective, consistent training is:
   a. Higher turnover.
   b. Fewer accidents.
   c. Better quality.
   d. B and c.
3. The first step in creating a training plan is to:
   a. Establish the standards of employee performance.
   b. Conduct training.
   c. Provide training moments.
   d. Evaluate the outcome of training.
4. What, who, how, when, and where are questions that need to be answered in which of the following?
   a. Deciding on what training materials are needed.
   b. Preparing to train employees.
   c. Evaluating training activities.
   d. All of the above.
5. Location of training is primarily based on:
   a. Transfer of learning.
   b. Employee attitudes.
   c. What type of subject you are trying to communicate.
   d. Who is doing the training.
6. A goal(s) in a great training session is:
   a. Make it difficult.
   b. Make it fun.
   c. Don't use visual aids.
   d. Make it passive.

7. To ensure the transfer of learning, one thing to consider is:
   a. Providing follow-up discussions.
   b. Catching people doing it wrong.
   c. Serving refreshments in class.
   d. Provide training at the end of the shift.
8. Informal role modeling/coaching training is known as
   a. The Discipline Process.
   b. Group training meetings.
   c. Training moments.
   d. Using visual aids.

Answers to the multiple choice questions are provided in Appendix A.

References/Suggested Readings

Zemke, R., Raines, C., Filipzpak, R. (2000), *Generations At Work. American Management Association.* New York, NY.

Suggested Web Sites

Gateway to Government Food Safety Information
www.foodsafety.gov

United States Department of Agriculture (USDA)
www.usda.gov

Centers for Disease Control and Prevention (CDC)
www.cdc.gov

United States Food and Drug Administration (FDA)
www.fda.gov

USDA Food Safety and Inspection Service (FSIS)
www.fsis.usda.gov

USDA/FDA Food and Nutrition Information Center
www.nal.usda.gov/fnic/

United States Environmental Protection Agency (EPA)
www.epa.gov

Partnership for Food Safety Education
www.fightbac.org

The Food Marketing Institute
www.fmi.org

Prentice Hall
www.prenhall.com

10 Food Safety Management Programs

Salmonella Outbreak Linked to Rotisserie Chicken from Local Deli

One hundred and twelve employees who attended a company picnic became ill with salmonellosis after eating the contents of boxed lunches purchased at a local retail food establishment. The most common symptoms experienced by the victims were diarrhea, cramps, bodyaches, chills, nausea, and fever. Employees at the establishment prepared the boxed lunches specifically for the picnic. It was a larger order than the facility was used to filling, but it was a great opportunity to add a new client. Inspectors from the local regulatory agency came to investigate. They asked the food employees to describe the food production practices used to prepare the suspect meals. One of the employees indicated filling the order for the boxed lunches was difficult, perhaps more than they could handle. Everyone was so rushed and the rotisserie ovens were overfilled with chickens. There was just no time to measure the temperature of the chickens during cooking. The employee indicated many of the chickens had blood and juices coming from them. This was very unusual, but the employees indicated they were not sure if this posed a problem.

Learning Objectives

After reading this chapter, you should be able to:

- ▲ Recognize how food safety management programs in retail food establishments reduce the risk of foodborne illness.
- ▲ Describe how to apply the Hazard Analysis Critical Control Point (HACCP) system for food safety management.
- ▲ Define Critical Control Points (CCPs) and identify critical limits as used in the HACCP system.
- ▲ Determine the roles of managers and employees related to food safety management.
- ▲ Discuss the need for a food recall procedure.
- ▲ Identify the need and function of product identification and product coding.

Essential Terms

Coding	**Good Retail Practices (GRPs)**
Critical Control Point (CCP)	**Hazard**
Critical limit	**Hazard Analysis Critical Control Point (HACCP) system**
Flow diagram	**Risk**
Food recall	**Standard Operating Procedures (SOPs)**

The Problem

The challenge of reducing the number of cases of foodborne illness has made it necessary to reevaluate our nation's food safety system. The retail food industry and food regulatory agencies must deal with:

- ▲ New hazards that cause foodborne illness
- ▲ A global food supply that carries more risks
- ▲ New techniques and equipment for processing, displaying, and serving food
- ▲ A change in eating patterns by customers including more ready-to-eat products

- An increase in the number of people who are highly susceptible to foodborne illness due to age and condition.

The size and scope of the food industry increases the workload on the FDA, USDA, and state and local food regulatory agencies. The FDA identified more than 30,000 food manufacturers and 20,000 food warehouses that process and store products. Inspection of all these operations is nearly impossible. The retail and foodservice portion of the food industry is even larger and consists of more than one million establishments with a work force of over 15 million people.

The number of foods imported into the United States increases every year. An increase in new methods used to process and prepare foods continues to grow. New items are produced and different foods are grown and imported. This challenges the government's ability to effectively monitor food safety. Governmental agencies at all levels have reduced staff and services due to higher costs and taxpayer mandates to decrease spending. These factors shift more responsibility for food safety to managers and employees that work in retail food establishments. Individuals who are engaged in food production and service must assume more responsibility to ensure the safety of the food they produce.

Food safety and food quality are both major concerns for retail food establishment managers and their staff. Food safety focuses on ensuring the products sold are safe to eat. Regulatory standards and guidelines related to safe food handling at retail levels are discussed in this book and in the *FDA Food Code*. Food quality is also very important. High quality often equals high profits and returning customers. With an average profit margin of 1-2%, it makes good business sense to produce and sell the safest and highest quality food possible.

The Solution

In Chapter 3 of this book, the major **risk** factors related to foodborne illness were identified. Recall that the most important factors were:

- Improper holding temperatures
- Poor personal hygiene
- Contaminated equipment
- Inadequate cooking
- Food from an unsafe source.

A food safety management program should protect foods from the factors known to lead to foodborne illness. To assure food safety, we need to focus on the following four keys to foodborne illness prevention:

- ▲ Time and temperature control
- ▲ Good personal hygiene practices and procedures
- ▲ Eliminate cross contamination
- ▲ An effective cleaning and sanitation program.

Other things you have learned in this book will also need to be part of your food safety system. For example, integrated pest management, facility and equipment design, a good crisis management program, and effective employee training must be incorporated into your food safety management program.

Food safety management programs in retail food establishments must be customized to fit company policies, the size of the establishment, and the number and types of customers served. To operate a food safety management program, you must understand the **hazards** that cause foodborne illness and how these hazards can be prevented. Your program must focus on eliminating or controlling the risk factors that are known to contribute to foodborne illness.

The food safety procedures in your management system must be easy for your employees to understand and implement. The way you manage this in your operation will be up to you. In this chapter, you will learn about some possible approaches that work.

Start with a Good Set of Instructions

The ability to do a job correctly depends on good instructions and properly trained employees. This holds true for all functions in a retail food establishment including marketing, purchasing, operations and customer services. This includes how to:

- ▲ Break down pieces of equipment for cleaning
- ▲ Stock and rotate products on shelves
- ▲ Wash hands properly
- ▲ Operate the slicer in the deli or meat department
- ▲ Take care of an emergency.

In retail food establishments, there are many different types of procedures that need to be done correctly to protect the safety of foods.

A detailed step-by-step form of instruction and training will help the employee do a job properly.

There are a wide variety of methods the retail food industry uses for this type of information. In some cases, a book of procedures is used. In other cases, procedures are posted near workstations for easy access by employees. There are a wide variety of terms that can be used for these procedures. "Best Practices" and "**Standard Operating Procedures**" (SOPs) are two terms commonly used to describe written procedures on how to perform a job. It does not matter what these procedures are called so long as they are written clearly, explained to employees, and performed correctly.

There are many different types of SOPs that can affect the safety and quality of foods during receiving, storage, preparation, and sale. Some examples of SOPs used by retail food establishments include:

- ▲ Good personal hygiene programs
- ▲ Cross contamination control
- ▲ Cleaning and sanitizing programs
- ▲ Pest management
- ▲ Storage practices of foods and ingredients
- ▲ Water and ice handling
- ▲ Use of chemicals.

The Food Marketing Institute has provided model SOPs for molluscan shellfish handling in a document titled "Molluscan Shellfish Handling for Retailers (FMI, 2001b)." In this document, six important elements are indicated for each SOP including:

- ▲ Title of the task to be completed
- ▲ Date the SOP was issued
- ▲ Area or department to use the SOP
- ▲ Detailed, specific actions that need to be taken to complete the task
- ▲ Any documentation that needs to be completed to verify the task was accomplished
- ▲ Approval signatures.

Figure 10.1 provides an example of a SOP from this document specific for personal health and hygiene practices. Note the SOP contains a very specific list of instructions and includes protocols for monitoring and corrective action. In this SOP, "critical items" are identified as those steps in the procedure most critical to food safety.

SOP for Personal Health and Hygiene

Background Notes **Date Effective 4/5/01**

- ▲ This procedure outlines personal health, hygiene and handwashing requirements required for the safe handling and sale of foods.
- ▲ Employees diagnosed with a food-related illness or showing symptoms of an illness are not permitted to work with food, food-contact equipment, utensils or surfaces
- ▲ Employees with festering cuts, abrasions, boils, or any condition that may cause flaking or exposed skin may not have contact with food or food-contact items.

Departments: All Departments

Specific Instructions

- ▲ **Critical Item – Only healthy employees are permitted to handle food, food-contact equipment, and utensils.** Employees that have been diagnosed with a food-related illness or are experiencing symptoms (sore throat with fever, diarrhea, vomiting or jaundice) must immediately notify their supervisor and remove themselves from the food environment.
- ▲ **Critical Item – Employees with festering cuts, abrasions, boils or any other condition that may cause flaking of exposed skin must immediately notify their supervisors and remove themselves from the food environment.** An infected wound must be freshly and properly bandaged and covered with an impermeable cover (surgical glove), which will prevent contact with food or food-contact items.
- ▲ Bathe daily and have clean, well-groomed hair. Present a clean appearance.
- ▲ Wear clean, washable outer garments, clean apron and clean shoes. Safety shoes may be necessary.
- ▲ Wear clean, effective hair restraint.
- ▲ Have clean hands and fingernails. False fingernails and nail polish are not permitted.
- ▲ Employees must avoid wearing any jewelry. Plain wedding bands and medical identification bracelets may be exempt.
- ▲ **Critical Item – All employees must wash their hands and exposed portions of their arms thoroughly:**
    a. Before starting work.
    b. After using the restroom.
    c. After eating, drinking, or using tobacco products.
    d. After touching their mouth or anything that has been in their mouth.
    e. After touching their hair, nose, ears, etc.
    f. After handling trash, garbage, dirty utensils or soiled equipment.
    g. After sneezing or coughing into their hands, or using a handkerchief or disposable tissue.
    h. After changing from raw to cooked or ready-to-eat foods.
    i. After any absence from the work area.

Figure 10.1 Model SOP for Personal Health and Hygiene for Molluscan Shellfish Handling

(Used with permission from the Food Marketing Institute – FMI, 2001b).

- ▲ **Critical Item – Washing hands and exposed portions of arms is accomplished by:**
  a. Vigorously rubbing hands together, lathering soap, using hot water for at least 20 seconds.
  b. Rinsing with clean water.
  c. Drying with disposable paper hand towels, clean continuous dispenser towels or air dryers.
- ▲ Avoid bare hand contact with any ready-to-eat foods, including shellfish. Use single-use gloves, deli tissues or utensils to handle product. Change single-use gloves frequently to prevent contamination. Remember gloves are not a substitute for proper hand washing.
- ▲ Remove personal items from pockets where they may fall into food, such as cigarettes, lighters and pens.
- ▲ Employees may only eat, drink, smoke or chew (tobacco or gum) in designated break areas.
- ▲ Employee street clothing, personal belongings or personally-owned foods and beverages should be stored in designated areas.

Monitoring

Daily: Department manager should visually observe employees for:

- ▲ Symptoms of illnesses or skin infections.
- ▲ Proper handwashing practices and frequency.

Corrective Action

- ▲ Refer all employees diagnosed with a food-related illness, displaying symptoms of illnesses or skin infection immediately to department manager.
- ▲ Refer employees to repeat handwashing procedures.

Figure 10.1 Model SOP for Personal Health and Hygiene for Molluscan Shellfish Handling (Cont.)

(Used with permission from the Food Marketing Institute – FMI, 2001b).

The HACCP Approach to Food Safety

In the food industry, there are a variety of programs that can be used to manage food safety. HACCP, or **Hazard Analysis Critical Control Point**, is a prevention-based program currently being used by many segments of the food industry to ensure food safety. Properly applied, the HACCP system controls hazards such as disease-causing microorganisms, chemical substances and physical objects that are common causes of foodborne illness (see Chapter 2).

The HACCP concept started in 1959 when the National Aeronautic and Space Agency (NASA) approached the Pillsbury Company to produce food safe for astronauts to eat in zero gravity conditions in the space capsules. In 1959, little was known about how foods might react under zero gravity. The most difficult and perhaps important aspect of the project was to develop a system that ensured nearly 100% freedom from contamination by microbial, chemical, or physical hazards. The system required control over all aspects of food production such as where raw material came from, processing, environmental conditions, personnel, storage, and distribution. This effort resulted in the beginning of the HACCP concept for food safety management.

The FDA and USDA now require HACCP programs for many different food manufacturing industries including meat and poultry slaughter and processing, seafood processors, and most recently, juice processors. HACCP is not required for retail food establishments except in a limited number of states and jurisdictions. However, many retail food establishments have implemented HACCP programs voluntarily to ensure the safety and wholesomeness of the products they produce and sell.

HACCP is commonly referred to as a systematic approach to ensure food safety. The focus of HACCP is to control the hazards that can cause foodborne illness by identifying when and where these hazards may occur and reducing the risk factors that contribute to foodborne illness during food production. HACCP programs will target potentially hazardous foods since they promote the growth of disease-causing microbes if mishandled. When a HACCP program is developed, a team of people is usually assembled to discuss two fundamental questions:

1. Are there any hazards or risks associated with the ingredients, the process, or the finished food?
2. How can I make sure the identified hazards and risks are kept at a safe level?

The Seven Principles in a HACCP System

The basic structure of a HACCP system consists of seven principles as listed in Figure 10.2. While each principle is unique, they all work together to form the basic structure of an effective food safety program. Before a HACCP system is developed, a team of people is usually assembled. The team's responsibility is to identify the potentially hazardous foods produced at the establishment and discuss the different processes employed to produce these items. After implementing the seven principles, effective training will be important to ensure that employees understand their roles and responsibilities related to HACCP.

"Hello! I'm Inspector Harry Hassup"

1. Hazard analysis.
2. Identify the critical control points (CCPs) in food preparation.
3. Establish critical limits (thresholds) which must be met at each identified CCP.
4. Establish procedures to monitor CCPs.
5. Establish the corrective action to be taken when monitoring indicates a critical limit has been exceeded.
6. Establish procedures to verify the HACCP system is working.
7. Establish effective record keeping that will document the HACCP system.

Figure 10.2 Seven Principles in a HACCP System
(Source: *FDA Food Code*)

Principle 1 – Hazard analysis. In this first step, the team identifies hazards associated with ingredients that are received, those that may be associated with each step in the flow of food, and those important at the time of purchase or consumption by the customer. Hazards are then assessed for their risk. Risk indicates a level of danger based on how likely the hazard is to occur and how severe the hazard could be from a public health standpoint. Those hazards most likely to occur or pose the highest risk of disease will be addressed in the HACCP system.

Principle 2 – Identify critical control points or CCPs. The *FDA Food Code* defines a CCP as a point in which loss of control may result in an unacceptable health risk. A CCP can be thought of as a point in the flow of food where a preventative measure can be put into place to control the identified hazards. The team identifies areas or points in the flow of a food product that must be controlled. These are the CCPs.

Principle 3 – Establish a critical limit for each CCP. **Critical limits** are the upper and lower boundaries of food safety and are set at each CCP to reduce hazards and risks to safe levels and assure the CCP is under control. Critical limits are usually based on time, temperature, pH, and moisture content of a food. When you set a critical limit, the goal is to prevent or eliminate the hazards identified during the hazard analysis portion of the program. A critical limit should be as specific as possible, such as "ground beef must be heated to an internal temperature of 155°F (68°C) or above for at least 15 seconds." A well-defined critical limit makes it clear and easy to determine when the limit has not been met. In retail HACCP plans, critical limits for time and temperature are most often used to keep hazards under control.

Principle 4 – Monitoring CCPs. Critical control points and critical limits are only effective if they are monitored during the flow of food. Monitoring includes activities such as measuring temperature and time. Monitoring helps to ensure the processes involved in the flow of food are done correctly and the hazards are in control.

Principle 5 – Take corrective action. Whenever the critical limit for a CCP is exceeded, something must be done to ensure the food is safe. When a critical limit is not met, fix it. Corrective action may include adjusting product temperature or time, or even discarding the product if necessary to ensure the safety of your customers.

Principle 6 – Establish procedures to verify the HACCP system is working. Once the HACCP system is in place, a designated process is used to make sure it is effective. The *FDA Food Code* recommends three phases to verification including 1) scientific verification of CCPs and CCP limits, 2) ensuring the HACCP plan is functioning effectively, and 3) repeated validation of the HACCP plan by internal or external food safety audits.

Principle 7 – Establish effective record keeping. This principle enforces the importance of documentation of CCPs, CCP limits, and monitoring activities. These records are the only proof the process is in control and the retail food establishment is complying with their HACCP plan.

Retail HACCP: A Process Approach

As the food industry develops and implements HACCP programs, it makes choices between a HACCP program that is "product" specific or one that is "process" specific.

- A product - specific HACCP program is tailored to the production of a specific food product. When product - specific HACCP programs are used, a HACCP plan is developed for each food product produced. This approach works well, especially in food manufacturing plants, where a limited number of products are handled.
- A process - specific HACCP program is customized to the processes used to produce a particular food or group of similar foods. In the retail food industry, numerous types of potentially hazardous foods are produced and sold within a single retail food establishment. It

often does not make sense and is not manageable to develop a HACCP plan for each one of these food products.

The "process approach" to HACCP simplifies the procedure so HACCP plan development is manageable. For example, for hot-holding, it does not matter what type of potentially hazardous food we are monitoring. In all situations, the key is to keep the food temperature above 140°F (60°C). The product is taken into consideration only when there is a difference in temperature guidelines such as cooking [i.e., ground beef must be cooked to at least 155°F (68°C) while poultry must be cooked to at least 165°F (74°C)].

Every process in the flow of food is created from a few basic steps. They may include:

- Received and store frozen
- Thaw
- Received cold
- Stored cold
- Prepare cold
- Display cold
- Cook
- Hot-hold
- Display hot
- Cool
- Reheat
- Service and handling.

If we take the process approach, we know there are certain temperature recommendations for each of these steps. We will follow these temperature guidelines in our food safety management system.

Figure 10.3 shows the different areas in the flow of food and the temperature guidelines used. These guidelines provide the basis for developing safety limits for the HACCP system.

Flow of Food Step	Temperature Guidelines for Potentially Hazardous Food (PHFs)
Receive frozen	Foods are solidly frozen when received
Frozen storage	Keep all foods solidly frozen
Thaw	Thaw in the refrigerator at 41°F (5°C), as part of the cooking process, under cool running water, or in a microwave oven
Receive cold	▲ Receive all Potentially Hazardous Foods (PHF) at 41°F (5°C) or below ▲ Whole shell egg may be received at 45°F (7°C) ambient temperature or below ▲ Dairy products may be received at 45°F (7°C) or below ▲ Shucked shellfish may be received at 45°F (7°C) or below ▲ Live shellstock may be received at 50°F (10°C) or below ▲ Reduced Oxygen Packaged (ROP) foods may have special temperature requirements
Refrigerated Storage Cold-Holding/ Display	▲ Store all PHFs foods at 41°F (5°C) or below ▲ ROP foods may have special temperature requirements
Prepare cold	Minimize time between 41°F (5°C) and 140°F (60°C) to less than 4 hours for PHF

Figure 10.3 Temperature Guidelines for Potentially Hazardous Foods During the Flow of Food

(Source: *FDA Food Code*)

Cooking	Cook to the safe product temperature ▲ Poultry and stuffed meats – 165°F (74°C) or above for 15 seconds ▲ Ground beef and pork and game animals – 155°F (68°C) or above for 15 seconds ▲ Shell eggs, beef and pork (other than roasts) and fish – 145°F (63°C) or above for 15 seconds ▲ Beef roast (medium), pork roast, and ham – 145°F (63°C) or above for 3 minutes ▲ All other PHFs – 145°F (63°C) or above ▲ Microwave cooking – 165°F (74°C) or above Allow to stand for two minutes after cooking
Hot-holding/ display	Hold all PHFs at 140°F (60°C) or above
Cooling	Cool all foods from 140°F (60°C) to 70°F (21°C) within 2 hours and from 140°F (60°C) to 41°F (5°C) within 6 hours (total cooling time should not exceed 6 hours from 140°F (60°C) to 41°F (5°C))
Reheat	Reheat all PHFs to greater than 165°F (74°C) within 2 hours
Preparation/ Handling/ Service	Restrict the amount of time potentially hazardous foods are held between 41°F (5°C) and 140°F (60°C) to 4 hours or less

Figure 10.3 Temperature Guidelines for Potentially Hazardous Foods During the Flow of Food

(Source: *FDA Food Code*)

Regardless of the type of food handled and sold in a retail food establishment, the food flow is made up of a series of the steps listed in Figures 10.3. For most of the steps in the flow of food, time and temperature control is the same regardless of the potentially hazardous food. There are a few exceptions. In receiving, eggs and Reduced Oxygen Packaging (ROP) foods may be received at a temperature other than 41°F (5°C). ROP foods may also have to be held cold at a temperature other than 41°F (5°C). Safe cooking temperatures will also be different depending on the product being cooked.

HACCP plans are often shown as flow diagrams. A flow diagram provides a visual map of the different ingredients and each of the steps involved in the handling and preparation of food. Flow diagrams may also include the CCPs included in the HACCP plan. Figures 10.4 and 10.5 provide examples of two different flow diagrams. One diagram is for a cooked and hot-held product (Figure 10.4), and the other is for a prepared ready-to-eat food displayed cold (Figure 10.5).

For the rotisserie chicken, cooking and hot-holding are identified as Critical Control Points (CCP). The critical limits for this product are cooking the chicken to an internal temperature of at least 165°F (74°C) for 15 seconds and holding it at 140°F (60°C) or above. Cooking is important for destroying any harmful bacteria that may have been present on the raw chicken. Control of hot-holding prevents growth of harmful bacteria that may have survived cooking or that contaminated the chicken after cooking.

A. Rotisserie Chicken

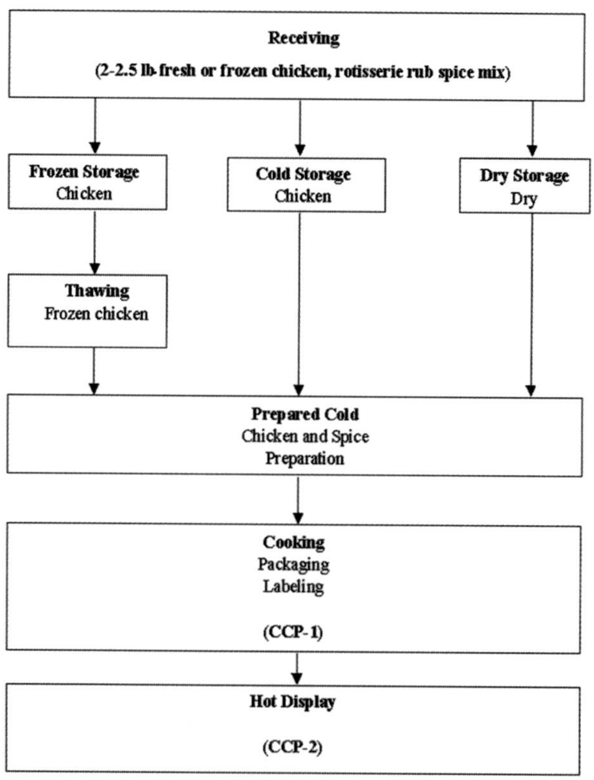

Figure 10.4 Example of Rotisserie Chicken Flow Diagram with CCPs Identified

For the RTE ham and cheese sandwich, cold storage, and cold-display are identified as a CCPs. The critical limits for cold storage and cold display are maintaining the sandwich temperature at 41°F (5°C) or below. Cold temperature control restricts the growth of harmful bacteria that could be present. During the preparation of the sandwich, preparation time [where foods are above may 41°F (5°C)] is best covered with GRP's. We know that foods can only be in the temperature danger zone for a maximum of 4 hours. In accordance, preparation time is usually limited to 2 hours or less to account for the prepared sandwich to cool down to below 41°F (5°C). In reality, however, preparation time for sandwiches is often limited to less than 30 minutes to maintain freshness and quality. Here again, the objective is to keep foods out of the temperature danger zone, so, minimizing preparation time should be the goal.

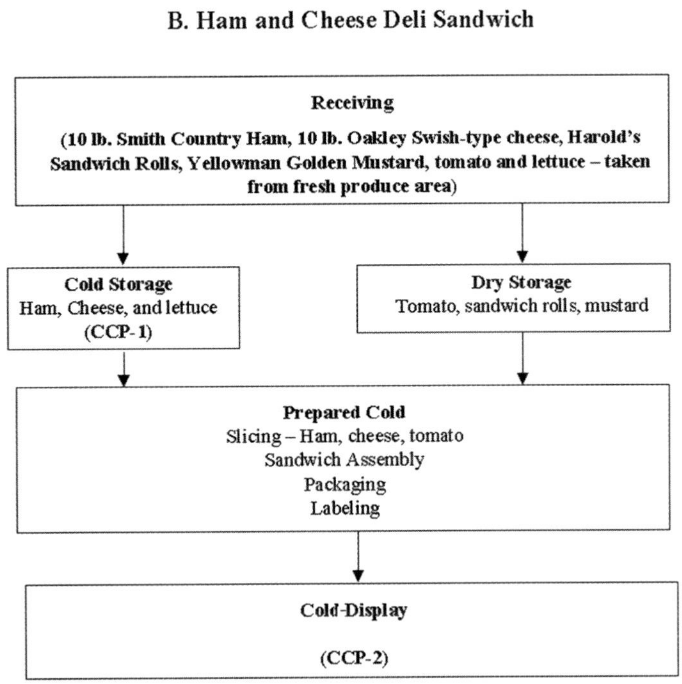

Figure 10.5 Example of Ham and Cheese Deli Sandwich Flow Diagram with CCPs Identified

Food Safety Management and HACCP Requirements

A growing number of state and local jurisdictions require retail food establishments to have a HACCP system in place. These agencies also conduct inspections of retail food establishments using HACCP principles. You may contact your local jurisdiction to obtain information about developing a HACCP system for your establishment.

The federal government requires HACCP plans and inspections for food processors that manufacture certain varieties of meat, poultry, and seafood products and unpasteurized juice products. A HACCP program is recommended when reduced oxygen packaged (ROP) foods are produced in retail food establishments. This HACCP system is required to prevent the growth of *Clostridium botulinum* in reduced oxygen packaged foods, smoked or cured vacuum packed foods, and garlic-in-oil mixtures. Other food products where HACCP should be implemented include molluscan shellfish and shellstock identification procedures.

Food safety management programs should be taken very seriously by the retail food establishment. The *FDA Food Code* recommends the person in charge have knowledge of:

- ▲ Foodborne illness prevention
- ▲ Application of the HACCP principles for food protection
- ▲ Requirements of the *FDA Food Code*.

If HACCP plans are put in place, the person in charge should be capable of effectively training food employees. In turn, food employees should be able to perform their responsibilities related to HACCP. These generally include monitoring, keeping records, and corrective actions.

Good Retail Practices

HACCP can be a very effective program but it is not the "cure-all" for all food safety problems and in all segments of the food industry. There are many other types of approaches that can be used.

In the food processing industry, a program called Good Manufacturing Practices (GMPs) is used as part of the food safety management program. GMPs refer to minimum sanitary and processing requirements necessary to ensure the production of safe and wholesome food. GMPs include provisions for food handling related to personnel, building and facilities, equipment and utensils, and production and process controls. When implemented properly, GMPs help control the possibility of contamination from poor personal hygiene, from chemicals, and from physical objects. They also help ensure cleaning and sanitizing are done properly. There are FDA requirements for GMPs in food manufacturing operations, and they are listed in Section 21 of the Code of Federal Regulations, Part 110.

More recently, on-farm production of foods has adopted a new term called Good Agricultural Practices, or GAPs. The goal of GAPs is to reduce the risk of contamination on farms mainly for fruit and vegetable products. Programs that use GAPs focus on farm management programs such as personal hygiene, irrigation water, manure, and cleaning procedures for freshly harvested produce. Ensuring your produce vendors adopt GAP type programs will be very important for reducing the risk of ready-to-eat produce displayed and sold in your establishment.

Following along the same pathway, the Food Marketing Institute has embraced a program called Good Retail Practices, or GRPs. **GRPs** are similar to the GMPs and GAPs used in food manufacturing and food production, respectively. The main difference is GRPs focus on food safety practices for retail food distribution, preparation, display and sale. In many retail food establishments, GRPs are considered an important and very effective way to manage food safety. GRPs can be flexible and written to enhance product quality and shelf life. Some examples of GRPs include:

- Receiving guidelines
- Storage guidelines
- Preparation guidelines
- Packaging and labeling guidelines
- Display guidelines
- Personal health and hygiene
- Employee and customer education
- Record keeping.

The Food Marketing Institute has published a document on sushi preparation called "FMI's recommendations for the safe preparation of sushi (FMI 2001a)." In this document, several examples of GRPs are provided. Examples of receiving and storage GRPs, respectively, from this document are provided in Figure 10.6.

A. Receiving Raw Sushi-Grade Fish

All ingredients should come from approved sources. Seafood products identified as a parasite hazard in the Seafood Hazard Guide should be obtained from a supplier who has frozen the product throughout to -4°F (-20°C) or below for seven days, or –31°F (-35°C) or below for 15 hours. If the species is identified as a histamine producer, documentation from the supplier should show temperature control from harvest through delivery was sufficient to prevent any stage of decomposition.

General Receiving Instructions:

- ▲ Check the incoming product for date code, physical condition of packaging, physical defects, insect infestation and food quality.
- ▲ Check the temperature to ensure the product is frozen solid or has been refrigerated to 41°F (5°C) or below.
- ▲ Place ready-to-eat foods in the cooler immediately.
- ▲ Place frozen product delivery into the freezer immediately.
- ▲ Rotate delivery of the product to facilitate the first-in first-out method of stock rotation.
- ▲ Obtain a letter of guarantee from the supplier stating all sushi-grade fish has been frozen at –4°F (-20°C) or below for at least seven days, or -31°F (-35°C) or below for 15 hours before it is released for distribution.
- ▲ Obtain a letter from the supplier verifying that, based on microbiological testing, the fish is high sushi-grade.
- ▲ Reject products that do not meet the above criteria.

B. Storage Guidelines for Raw Sushi-Grade Fish

To maintain product quality once the retailer has accepted shipment:

- ▲ Place product in the appropriate storage unit.
- ▲ Equip refrigeration unit with a calibrated thermometer easily accessible for viewing.
- ▲ Use recording devices and alarm systems.
- ▲ Avoid cross contamination.
- ▲ Store product in a specific, pre-designed area in the refrigeration unit, away from other raw and cooked foods.
- ▲ Keep product containers 6 inches above the floor, away from walls and ceilings, and preferably on non-wooden supports, shelving, or surfaces.
- ▲ Use an inventory system to make sure the first product in is the first product out (FIFO).
- ▲ Check the temperature of the refrigerator and freezer units at least twice daily to ensure they are operating at 0°F (-18°C) and 41°F (5°C) or below respectively.
- ▲ Notify store manager or person in charge immediately if improper temperatures are detected at times other than the unit defrost times. Remove food from defective unit and place it in one that works properly.
- ▲ Store raw foods below or away from ready-to-eat foods.

Figure 10.6 GRPs for the Safe Preparation of Sushi for a) Receiving and b) Storage

(Used with permission from the Food Marketing Institute – FMI, 2001a).

Food Recall Procedures

A retail food establishment has limited control over the food they sell. It makes good sense to ensure you have checked out your supplier to determine you are receiving the best products possible. Once in the store, your job is to follow good food-handling practices to keep products safe and in good quality. Once in a while, however, contaminated products may be received from food manufacturers, slaughterhouses, or farms that can present a hazard to your customers. In these situations, the company that supplied you with contaminated product may put a "food recall" in place. If this happens, your job is to remove the contaminated product from your shelves so your customers cannot purchase it. Another important part of your food safety management system is to develop a food recall procedure.

A **food recall** is the system used to protect customers from using a product that may be unsafe. There are different types of recalls that can be issued depending on how much of a risk the product may present. Chapter 11 describes these different levels of food recalls. Most of the time, a recall involves removing the product from retail shelves so it cannot be sold. It may also involve retrieving product that has already been sold or providing information to customers that have already purchased the product.

The purpose of a recall is to remove a food product from commerce when there is reason to believe it may be adulterated (injurious to health or unfit for human consumption) or misbranded (false or misleading labeling and/or packaging). There must be a procedure in place to successfully manage a food recall. Be well prepared. Develop specific recall policies and procedures and maintain communication with regulatory agencies and the public. In most cases, the manufacturer of the product initiates the food recall. Many times this is done with the assistance of a federal and/or local regulatory authority.

In the event of a food safety emergency or other emergencies involving security and loss prevention issues, FMI members can use the 24-hour Crisis Notification System to contact FMI staff after regular working hours.

Coding and Product Identification

When monitoring is done, especially for time and temperature, it is important that you be able to identify which product was actually measured. You need a procedure to identify the product. This is usually done with a process known as product **coding**.

Product coding will be different in almost every retail food establishment because it needs to be customized to your in-house operations. One of the more common uses of product coding is to ensure foods are displayed and used on a FIFO basis and that product with an expired shelf life is discarded. Figure 10.7 provides examples of a coding system that can be used in a retail food establishment's deli operation.

Challenges of Food Safety Management

There are a tremendous number of challenges related to managing food safety in retail food operations. In the past decade, there has been a debate over how retail food safety programs should be developed and implemented. Like many within the foodservice and food industry, the retail food industry has a high employee turnover rate. There are language and cultural barriers that make food safety education difficult. The retail food industry and regulatory community do not always agree on food safety practices. Even with all these challenges, retail food safety management must be a high priority.

As you develop a food safety management program for your establishment, use the following guidelines:

- Focus on risk: Inspection procedures and audits for food safety management programs should focus directly on controlling risk factors that affect food safety.
- Make it manageable: The food management program needs to be user friendly. It should be easy to implement and manage. Food safety management programs need to be flexible and related to specific operational procedures. One size does not fit all.
- Strengthen communication: The retail food industry and regulatory professionals share a common goal to protect public health by only allowing the sale of safe food. With the same goal, the food industry and regulatory professionals should partner and work together. Education and training need to be encouraged for members of the retail food industry and for regulators.

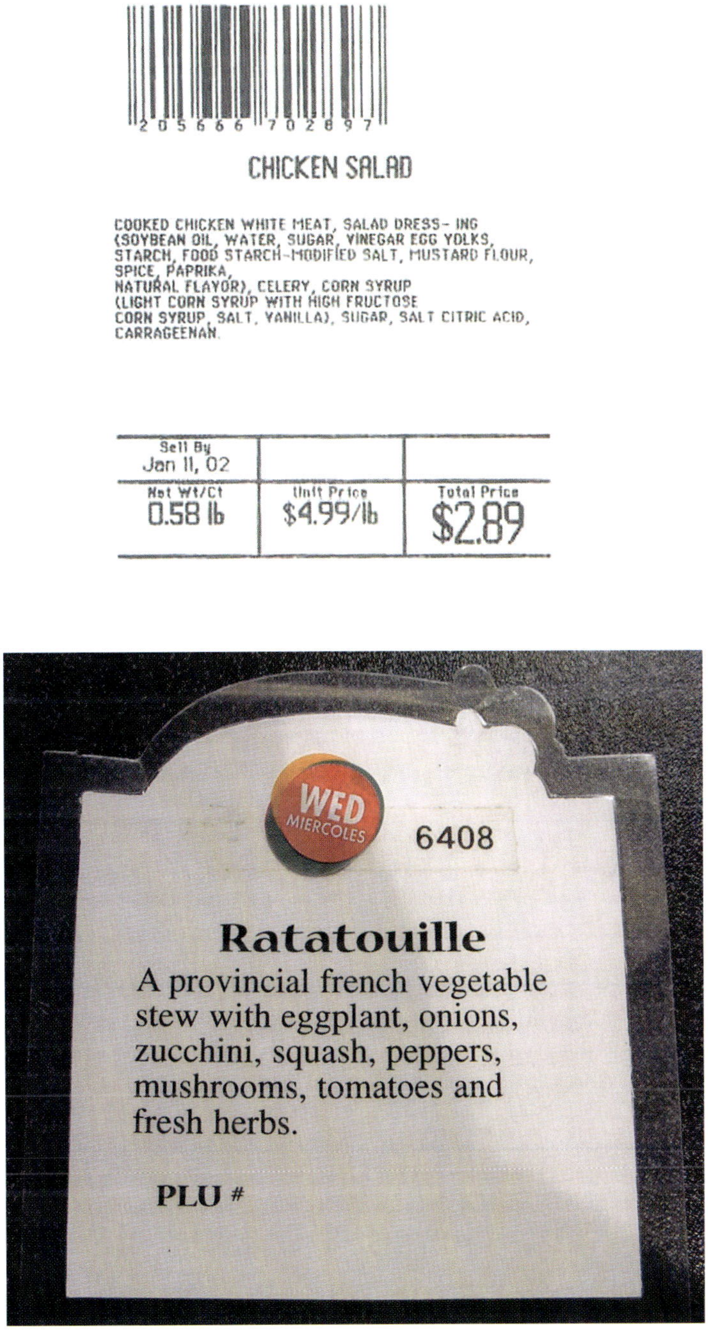

Figure 10.7 Coding Systems Used in Deli Operations

Summary

If you implement a food safety management system, you can avoid situations like the one described in the case at the beginning of this chapter. When a food safety management system is properly followed, it protects against many types of mishandling. Poultry has frequently been involved as the source of foodborne salmonellosis. If *Salmonella* bacteria are present after the cooking process, the bacteria can multiply to high levels. Chicken should be cooked to an internal temperature of 165°F (74°C) for 15 seconds (CCP-1) to kill the germs that may be present in and on the raw birds. Careful attention should also be placed on preventing cross contamination from equipment and a food employee's hands. Temperatures should be checked at the end of the cooking process to ensure 165°F (74°C) was obtained. In this case, the retail food establishment was trying to fill an order well above what it normally does. It should have compensated for an increased volume of food. Chickens could have been cooked in several batches or cooked longer to ensure safety. Employees must know how to monitor the final cook temperature of the chicken and record their measurements. Employees must be encouraged to contact their supervisor if they discover the prescribed critical limit is not being met. Management must then take the steps needed to correct the problem and bring the production process under control. Another viable food safety option would be to select an alternative food item for the picnic that presented lower risk and was more manageable to produce.

Discussion Questions (Short Answer)

1. What is HACCP?
2. What types of foods are incorporated into the HACCP system?
3. What hazards are incorporated into the HACCP system?
4. Define critical control point (CCP) and give some examples.
5. Define a critical limit and provide some examples.
6. What are SOPs? Provide some examples.
7. What are GRPs? Provide some examples.
8. Why is a procedure for food recalls important?

Quiz 10 (Multiple Choice)

1. A hazard as used in connection with a HACCP system is:
   a. Any biological, chemical, or physical property that can cause an unacceptable risk.
   b. Any single step at which contamination could occur.
   c. An estimate of the likely occurrence of a hazard.
   d. A point at which loss of control may result in an unacceptable health risk.
2. A risk as used in connection with a HACCP system is:
   a. Failure to meet a required critical limit for a critical control point.
   b. An estimate of the likely occurrence and severity of a hazard.
   c. Greatest for nonpotentially hazardous foods.
   d. Lowest when complex recipes are required to produce a food item.
3. Which statement about HACCP programs is false?
   a. The HACCP system attempts to anticipate problems before they happen and establish procedures to reduce the risk of foodborne illness.
   b. The HACCP system targets the production of potentially hazardous foods from start to finish.
   c. Records generated by the HACCP system can also be used to aid in foodborne disease investigations.
   d. A HACCP system should only be implemented by public health officials who have been certified by the FDA to conduct such programs.
4. Which of the following is an example of a common critical control point in a HACCP plan?
   a. Washing fruits and vegetables.
   b. Cooking hamburgers.
   c. Handwashing.
   d. Inspecting procedures for quality.

5. Which of the following is an example of a critical limit?
    a. Poultry and eggs are purchased from approved sources.
    b. Chicken and noodles are heated on the stove until the center of the product reaches 165°F (74°C) for 15 seconds.
    c. Only pasteurized milk is used by the establishment.
    d. The cutting board is washed and sanitized between chopping carrots and celery for the garden salad.
6. The ultimate success of a food safety management program depends on:
    a. Eliminating all potentially hazardous foods from your operation.
    b. Providing proper training and equipment for employees who are implementing the HACCP system.
    c. Having HACCP flow charts developed for every single food sold by the store.
    d. Grocery store managers having sole authority for implementing the HACCP system.
7. A food recall is:
    a. A procedure for removing unsafe foods from retail shelves.
    b. A process for labeling food products during preparation.
    c. Only important for very large retail food operations.
    d. Only important for USDA inspected foods.

Answers to the multiple choice questions are provided in Appendix A.

References/Suggested Readings

Bryan, Frank L. 1990. *Hazard Analysis Critical Control Point (HACCP) Systems for retail food and restaurant operations.* Journal of Food Protection 53(11): 978-983.

Bryan, F. L. 1992. *Hazard Analysis Critical Control Point Evaluations: A Guide To Identifying Hazards and Assessing Risks Associated with Food Preparation and Storage.* World Health Organization. Geneva, Switzerland.

Bryan, F.L., C.A. Bartleson, C.O. Cook, P. Fisher, J.J. Guzewich, B.J. Humm, R.C. Swanson, and E.C.D. Todd. 1991. *Procedures to Implement the Hazard Analysis Critical Control Point System.* International Association of Milk, Food and Environmental Sanitarians, Ames, IA, 72 pp.

Corlett, D.A. and M.D. Pierson. *HACCP Principles and Applications,* Van Nostrand Reinhold, New York, 1992

Federal Register, Volume 59, Number 149. U.S. Government Printing Office, 1994. Washington, DC.

Food and Drug Administration. 2001. *2001 Food Code.* United States Public Health Service. Washington, DC.

Food Marketing Institute. 2001a. *FMI's Recommendations for the Safe Preparation of Sushi.* Food Marketing Institute. Washington, DC.

Food Marketing Institute. 2001b. *Molluscan Shellfish Handling for Retailers – A Total Food Safety Management System Guide for Molluscan Shellfish.* Food Marketing Institute. Washington, DC.

LaVella, B. and J. L. Bostic. 1994. *HACCP for Food Service--Recipe Manual and Guide.* LaVella Food Specialists. St. Louis, MO.

J. K. Lotkin. 1995. *The HACCP Food Safety Manual.* Wiley Press. New York, NY.

Pierson, M. D. and D. A. Corlett, Jr. 1992. *HACCP Principles and Applications.* Van Nostrand Rheinhold. New York, NY.

Suggested Web Sites

Canadian Food Safety Sites Involving HACCP
foodnet.fic.ca/safety/safety.html

International Food Information Council
ificinfo.health.org/

NSF International
www.nsf.org

The Food Marketing Institute
www.fmi.org

Gateway to Government Food Safety Information
www.foodsafety.gov

United States Department of Agriculture (USDA)
www.usda.gov

Centers for Disease Control and Prevention (CDC)
www.cdc.gov

United States Food and Drug Administration (FDA)
www.fda.gov

USDA Foods Safety and Inspection Service (FSIS)
www.fsis.usda.gov

USDA/FDA Food and Nutrition Information Center
www.nal.usda.gov/fnic/

Partnerships for Food Safety Education
www.fightbac.org

Food and Agriculture Organization
www.fao.org

11 Food Safety Regulations

Infant Formula Recall

On July 1, 2001, 4.6 million cans of infant formula were recalled because the cans had incorrect preparation instructions in Spanish that could lead to serious health problems and even death (The Indianapolis Star, 2001). The formula is intended as the sole source of nutrition for infants younger than 6 months with digestive problems or who are allergic to proteins. Consumers were given an 800 number to call the company or use the instructions on the can label to return the formula.

How can retail food establishments assist in this mandatory recall?

Learning Objectives

After reading this chapter, you should be able to:

- ▲ Recognize the role of federal, state, and local jurisdictions in regulating and monitoring food safety.
- ▲ Identify the parts of retail food establishments included in inspections and audits.
- ▲ Use this chapter as a reference for federal, state, and local authorities.
- ▲ Identify organizations important to the retail food industry.

Essential Terms

Adulterated

Centers for Disease Control and Prevention (CDC)

Conference for Food Protection (CFP)

Generally Recognized as Safe (GRAS) substances

Grading

Inspection for wholesomeness

Jurisdiction

Misbranded

Recall of a food product

State and local departments of agriculture

State and local departments of health

Environmental Protection Agency (EPA)

Food and Drug Administration (FDA)

United States Department of Agriculture (USDA)

State and Local Regulations

State and local agencies responsible for enforcing food and safety requirements may be under the department of environmental health, public health, or agriculture. **Jurisdiction** varies from state to state. Contact your local authorities to find out who is in charge of food safety in your area.

Most retail food establishments will work closely with their local health department. This agency can provide you with a copy of its current food safety code. **Your state and local code contains the food safety standards and regulations that would apply to your type of operation.** Familiarize yourself with the key food safety provisions of the code.

Permit to Operate

It is unlawful to operate a retail food establishment unless you have a valid permit issued by the regulatory authority in your area (Figure 11.1). Apply for the permit well in advance of the anticipated date for opening the establishment. Many health agencies require submission and approval of blueprints and specifications for an establishment before new construction or extensive remodeling can begin. Contact your local health agency to find out what level of plan review is required in your community.

Retail Food Market

Monroe County Health Department
Bloomington, IN 47404-3989

Name: ____________________
Address: ____________________
City/ State ZIP: ____________________

Owner: ____________________

Having complied with the rules and regulations of the Monroe County Health Department as authorized by the Indiana Administrative Code and the Board of Commissioners of Monroe County, Indiana, is hereby authorized to operate a Retail Food Market at the above location for the calendar year ending December 31st:

Issued ____________________

By ____________________

2001

This License Is Not Transferable to Another Individual or Location

Figure 11.1 Permit to Operate

When a permit is obtained, post it in a prominent location where it can be seen. Permits are generally not transferable from one person to another, from one retail food establishment to another, or from one food operation to another. If you add a partner or incorporate, these actions are considered ownership changes and may necessitate application for a new permit to operate.

The number of inspections or audits that must be performed each year by representatives of food regulatory agencies is usually prescribed in the retail food code. However, many health departments are currently using a risk-based approach for setting inspection frequencies. Under this system, the risk a food establishment poses is calculated using such factors as:

- ▲ The sanitation history of the establishment
- ▲ Number of home meal replacements sold
- ▲ The amount of food preparation and handling conducted
- ▲ Condition of building
- ▲ Critical violations observed.

Low-risk facilities are usually inspected once a year or less. High-risk facilities, on the other hand, may be inspected four or more times per year. A risk-based inspection system enables local health departments to properly utilize their resources with those establishments that need them most.

Always cooperate with inspecting personnel. Greet the inspector when he or she arrives and ask to see their official identification. Whenever possible, accompany the inspector during an inspection or audit (Figure 11.2).

Figure 11.2 Inspection

A routine inspection or audit normally consists of three phases:

1. A pre-inspection conference where the inspector and manager or person in charge review previous inspection results and discuss information relevant to the current inspection.
2. The current inspection is conducted.
3. A post-inspection conference is conducted where the results of the current inspection are reviewed and discussed with the manager or person in charge.

Most inspections will focus on:

- ▲ Foods and supplies
- ▲ Personal hygiene and employee health
- ▲ Temperatures of food and food-holding equipment
- ▲ Cleaning and sanitizing procedures
- ▲ Equipment and utensils
- ▲ Water supply and waste disposal
- ▲ Pest control and other aspects of the operation that might compromise food safety.

If a critical violation such as temperature abuse or cross contamination is discovered during an inspection, corrective action must be taken immediately. Failure to do so could result in fines and possible closure of the establishment. Lesser violations should be corrected as soon as possible, but always prior to the next inspection or audit.

Food managers should periodically conduct self-inspections of their establishment to ensure proper sanitation and food safety between health department inspections. During a self-inspection, the manager will look for signs of contamination or conditions that might be hazardous to food safety.

Federal Agencies

The primary federal agencies that protect our food supply include the:

- ▲ Food and Drug Administration (FDA)
- ▲ United States Department of Agriculture (USDA)
- ▲ Environmental Protection Agency (EPA).

Other important government agencies and organizations include the:

- ▲ National Marine Fisheries Service (NMFS)
- ▲ Centers for Disease Control and Prevention (CDC)
- ▲ Occupational Safety and Health Administration (OSHA)
- ▲ Consumer Product Safety Commission (CPSC)
- ▲ Federal Trade Commission (FTC).

Most states have their own counterparts of the federal agriculture, and *health departments, OSHA, and EPA.*

Food and Drug Administration (FDA)

The **Food and Drug Administration (FDA)** is an agency of the U.S. Department of Health and Human Services. The FDA is recognized as the leading food and drug regulatory agency in the world. This agency regulates the processing, manufacturing, and interstate sale of many food items. All foods must comply with FDA food safety regulations EXCEPT domestic or wild cattle, sheep, swine, goat, horse and mule; and domestic chicken, turkey, duck, goose, guinea fowl, and egg products.

The FDA sets standards with respect to composition, quality, labeling, and safety of foods and food additives. The FDA's duties include performing routine audits of food processing plants, food storage, and imported foods to ensure safety standards are maintained. The agency is also responsible for the toxicological safety of food packaging materials. Additionally, the FDA seeks to protect the public's health by preventing the adulteration and misbranding of food.

The FDA's goal is to ensure food shipped in interstate commerce is safe, pure, wholesome, sanitary, and honestly packaged and labeled. The agency also helps to prevent milk-borne and shellfish diseases and assists in the control of sanitation related to shellfish production. The FDA is responsible for maintaining a list of interstate certified shellfish shippers and a list of interstate milk shippers. The lists indicate those operations that have satisfactorily met sanitation requirements of the national shellfish program and national milk sanitation program, respectively. These lists are made available for food control officials, the food industry, and other interested parties.

The FDA has published the *FDA Food Code* (Figure 11.3), which serves as a model for retail food programs regulated by federal, state, local, and tribal agencies. The *FDA Food Code* is not a federal law or federal regulation, and it does not replace existing retail food laws. *The Code* is more like a set of recommendations put forth to promote food safety and sanitation nationwide.

The FDA works closely with other federal, state, local and tribal food regulatory authorities in matters of food protection. The FDA and the food processing industry cooperate to provide a clean and wholesome national food supply.

The FDA can request a **recall of a food product** if the agency has determined a potential hazard exists. This would justify regulatory action should such action become necessary. Food recalls are "voluntary" with the possible exception of a recall involving infant formula. In the vast majority of cases, food processors voluntarily remove products from the marketplace to keep consumers safe. However, if a firm does not agree with an FDA request for a voluntary recall, the agency may issue public press releases and proceed to seize the product in order to remove it from commerce.

Figure II.3 2001 *FDA Food Code*

Regulators and industry personnel agree a properly managed voluntary recall is the most rapid, effective, and efficient means of achieving product removal. Three types of food recalls, Class I, Class II, or Class III, can be issued depending on the severity of the health risk (Figure 11.4). It is important all food retail establishments have a policy regarding the handling of food recalls. Recalled foods should be promptly removed from retail shelves and cannot be sold.

Class I	Foods that may cause serious adverse health consequences
Class II	Foods that would result in a temporary or reversible health problem
Class III	Foods that are not likely to cause danger to health

Figure II.4 Food Recalls

The person in charge of a retail food establishment must assure foods involved in a recall are promptly removed from retail shelves or from service. This prevents customers from purchasing and consuming potentially unsafe foods. If possible, the crises management team of retail food establishments should contact customers who have already purchased a food that has been recalled. Some retail food establishments are able to contact customers through the use of "value added" cards scanned during checkout. Retail food establishments can also use local radio, television, and newspapers, and post notices inside the store to notify customers about food products being recalled.

U.S. Department of Agriculture (USDA)

The **United States Department of Agriculture (USDA)** has many agencies engaged in food protection. The Agricultural Marketing Service (AMS) conducts programs that aid in the marketing of agricultural products. The AMS develops quality standards and offers voluntary grading services for red meats, poultry and eggs, dairy products, and fruits and vegetables.

Many types of quality standards have been developed to help ensure food quality and define differences in quality that affect the usefulness and value of a food. Grade or quality standards cover the whole range of a product's natural quality. The number of grades for a particular food depends on its variability. For instance, eight grades are necessary to cover the range of beef quality, whereas only three are needed for poultry. Beef grades usually encountered are USDA Prime and Choice. Choice is the grade most widely available; it denotes a high degree of tenderness, juiciness, and flavor. USDA Grade A poultry has more meat and a better appearance. Lower grades, such as B and C, are seldom offered for retail sale. More information on grading is contained in Chapter 4 of this book and examples of inspection and grading stamps for meat, poultry, and eggs can be found in Figures 4.6, 4.7, and 4.9, respectively.

The Food Safety and Inspection Service (FSIS) of the USDA inspects domestic and imported meats, poultry, and processed meat and poultry products. Employees of the FSIS inspect meat in accordance with the *Federal Meat Inspection Act of 1906* to ensure a clean, wholesome, disease-free meat supply. USDA maintains a list of approved facilities for meat and poultry processing. Inspections are made by veterinarians or persons under their supervision at places of animal slaughter and at meat processing facilities. Poultry is inspected under the provisions of the *Federal Poultry Products Inspection Act of 1957*. This law requires poultry to be processed in plants having full-time government inspection service. Poultry is inspected live prior to slaughter, during evisceration (removal of the internal organs), and during or after packaging. The *Egg Products Inspection Act of 1970* provides for the inspection of egg product processing plants. It also provides for a surveillance program to help ensure eggs still in the shell and that are unfit for human consumption (incubator rejects, leakers, etc.) are being properly restricted or disposed of by egg handlers.

Purchasers should be aware of the difference between inspection and grading. **Inspection for wholesomeness** refers to an official examination of food to determine whether it is wholesome and free from adulteration. **Grading** refers to the process of evaluating foods relative to specific, defined standards in order to assess its quality. Food laws require inspections for wholesomeness, but grading is optional. Most purchasers prefer to buy graded products because they know those products have met specific quality standards. Of the two services, inspection for wholesomeness is the most important for food safety.

USDA can also be involved with microbial testing of foods. Because of the associated risks, *E. coli* O157:H7 has been declared an adulterant in raw ground beef. Under this provision, there is legally no tolerance (zero tolerance) for *E. coli* O157:H7 in ground beef. A national microbial sampling plan was initiated to understand the prevalence of *E. coli* O157:H7 in ground beef in an effort to develop better control methods. Under the sampling plan, USDA officials take ground beef samples from federally inspected plants and in retail stores and test for *E. coli* O157:H7. If a sample is confirmed positive for *E. coli* 0157:H7, the food establishment may be asked to recall the product and communicate this information to the public. It is important to develop a written procedure (i.e., SOP) for ground beef sampling and that measures are in place to identify different lots of ground beef.

U.S. Department of Commerce (USDC)

Grade standards for processed fishery products have been established by the Voluntary Inspection Service of the U.S. Department of Commerce. The Department's National Marine Fisheries Service (NMFS) administers these programs on a voluntary, fee-for-service basis. They also maintain a list of approved fish processors and fishery products. A fish processor may decide to have an inspector stationed in his plant on a permanent basis to assess the premises for sanitary conditions, determine wholesome ingredients are used, and grade the product. This type of inspection is strictly voluntary. The product gets evaluated and graded as A, B, or C in quality, the last grade being the lowest acceptable quality. Examples of NMFS-regulated products include frozen raw and fried breaded fish portions, shrimp, frozen fried scallops, and frozen sole and haddock fillets.

Environmental Protection Agency (EPA)

The **Environmental Protection Agency (EPA)** was established in 1970. It was created to prevent, control, and reduce air, land, and water pollution. Related to food safety, the EPA regulates the use of toxic substances (pesticides, sanitizers, and other chemicals), monitors compliance, and provides technical assistance to states.

The EPA regulates pesticides under two major federal statutes. Under the Federal Insecticide, Fungicide, and Rodenticide Act (FIFRA), the EPA registers pesticides for use in the United States and prescribes labeling and other regulatory requirements to prevent unreasonable adverse effects on health or the environment. Under the Federal Food, Drug, and Cosmetic Act (FFDCA), the EPA establishes tolerances (maximum legally permissible levels) for pesticide residues in food. Tolerances are enforced by the Food and Drug Administration (FDA) for most foods, and by the U.S. Department of Agriculture/Food Safety and Inspection Service (USDA/FSIS) for meat, poultry, and some egg products.

The EPA is also mainly responsible for the implementation of 10 major environmental laws. These laws cover an enormous range of environmental problems including air pollution, water pollution, hazardous wastes, noise pollution, and contaminated drinking water. Some of these laws, especially those controlling the handling and disposal of solid and hazardous wastes, may affect retail food establishment operations. The establishment manager should contact the state and local environmental agencies to be sure their facility is in compliance with current regulations.

Centers for Disease Control and Prevention (CDC)

The **Centers for Disease Control and Prevention (CDC)** is the federal agency responsible for protecting public health through the prevention and control of diseases. The CDC supports foodborne disease investigation and prepares annual summaries and statistics on outbreaks of foodborne diseases including all those transmitted through food and water.

Occupational Safety and Health Administration (OSHA)

The **Occupational Safety and Health Administration (OSHA)** was created in the Department of Labor to enforce the Occupational Safety and Health Act. It enforces health standards and regulations on safety, noise, and other workplace-related hazards to protect employees from illness and injury when they are on the job. The agency conducts investigations and inspections to assess health and safety conditions in places of employment.

Federal Trade Commission (FTC)

The **Federal Trade Commission (FTC)** is involved in enforcing various laws regarding marketing practices and national advertising of foods and other products.

Other National Food Safety Related Organizations

The **Conference for Food Protection (CFP)** is a nonprofit organization that represents the major stakeholders in retail food safety. Its membership includes representatives of the food industry, government, academic community, professional organizations, and consumer groups. The mission of the CFP is to promote food safety and consumer protection. It meets at least every other year to identify food safety problems, make recommendations, and implement practices to ensure food safety. The CFP has no formal regulatory authority, yet over the years it has been able to effectively influence the content of model retail food laws and regulations.

Other organizations also help to protect our food supply. Many of the following organizations have good information and training materials:

- ▲ The American Public Health Association (APHA)
- ▲ The Association of Food and Drug Officials (AFDO)
- ▲ The Food Marketing Institute (FMI)
- ▲ The Frozen Food Industry Coordinating Committee (FFICC)
- ▲ The Food Research Institute (FRI)
- ▲ Institute of Food Technologists (IFT)
- ▲ National Conference on Interstate Milk Shipments (NCIMS)

- International Association for Food Protection [formerly the International Association of Milk, Food, and Environmental Sanitarians (IAMFES)]
- The National Environmental Health Association (NEHA)
- The National Pest Management Association (NPMA)
- The National Restaurant Association (NRA)
- NSF International (NSF)
- The National Shellfish Sanitation Program (NSSP)
- Underwriters Laboratories Inc. (UL)
- World Health Organization (WHO).

Food Law

Federal Food, Drug, and Cosmetic (FD&C) Act

The FDA administers the federal *Food, Drug, and Cosmetic Act* originally passed by Congress in 1938 to ensure foods are safe, pure, wholesome, and produced under sanitary conditions. Several amendments have been added to the law since then. Among the most notable amendments are the *Miller Pesticide Amendment of 1954*, the *Food Additives Amendment of 1958*, and the *Color Additives Amendment of 1960.*

The *Miller Pesticide Amendment* establishes a procedure for the setting of safe amounts (tolerances) for residues of pesticides that may remain on fresh fruits, vegetables, and other raw agricultural commodities when shipped in interstate commerce.

The *1958 Food Additives Amendment* has a two-fold purpose: (1) to protect the public health by requiring proof of safety before a substance may be added to food, and (2) to advance food technology and improve the food supply by permitting the use of substances which are safe at the levels of intended use. Before something is approved for use by the FDA, data must be supplied proving the substance is free from harm. If approved, the FDA will then define limits that can be used in foods. Currently, there is a list of more than 3,000 approved food additives. Excluded are those food additives that have been designated as **GRAS** or **generally recognized as safe** substances. These are additives that have been used in foods for years and with no apparent ill effects. Substances that typically fall into the GRAS category are spices, natural seasonings, flavoring materials, fruit and beverage acids, and baking powder chemicals. The purpose of establishing a GRAS list is to recognize the safety of basic substances without the requirement for rigorous safety testing.

The *Color Additive Amendments of 1960* were enacted to control the use of natural and synthetic color additives. Color additives are substances added only for the color they give. Color additives must be safe under normal conditions of use and must not cause cancer when ingested by humans.

1906—Federal Meat Inspection Act

This law provides for the mandatory inspection of animals, slaughtering conditions, and meat processing facilities. It helps to ensure meat and products containing meat are clean, wholesome, free from disease, and properly represented. Such products are stamped "U.S. Inspected and Passed by Department of Agriculture." This stamp is necessary for interstate commerce and imported goods. This law is enforced by the Food Safety Inspection Service of the Department of Agriculture.

For products not crossing state lines, the Wholesome Meat Act is enforced (usually governed by state health departments).

1957—Federal Poultry Products Inspection Act

This law provides for the mandatory inspection of poultry, slaughtering conditions, and poultry processing facilities. It helps to ensure poultry and products containing poultry are clean, wholesome, free from disease, and properly represented.

1970—Egg Products Inspection Act

Under the provisions of this act, the Food Safety and Inspection Service (FSIS) inspects facilities that process egg products. These are facilities that break, pasteurize, freeze, dry, and package egg products. The Egg Products Inspection Act also authorizes FSIS or its designee to provide a surveillance program to help ensure eggs still in the shell and that are unfit for human consumption (incubator rejects, leakers, etc.) are being properly restricted or disposed of by egg handlers.

1996—Food Quality Protection Act

Congress unanimously passed landmark pesticide food safety legislation known as the Food Quality Protection Act (FQPA). The FQPA mandates a single, health-based standard for all pesticides in all foods;

provides special protection for infants and children; expedites approval of safer pesticides; and requires periodic reevaluation of pesticide registrations and tolerances to ensure the scientific data supporting pesticide registrations remain up to date.

Food Labeling

Nutrition and Ingredient Labeling

The USDA and the FDA share responsibility for enforcing the Nutrition Labeling and Education Act. This law provides information on specific nutritional guidelines and defines what needs to be on a Nutrition Label (Fig. 11.5). This label, which provides information for processed foods on protein, fat, carbohydrate, and mineral content, helps customers make informed choices when selecting foods.

Meltaway/French Mint Combo 5 1/2 oz.

Nutrition Facts

Serving Size: 42.0g, 5 pieces
Serving Per Container: 4

Amount Per Serving

Calories 248 Calories from Fat 150

	% Daily Value*
Total Fat 17g	**26%**
Saturated Fat 10g	**50%**
Cholesterol Less then 5mg	**0%**
Sodium 50mg	**2%**
Total Carbohydrate 22g	**7%**
Dietary Fibers 1g	**4%**
Sugars 20g	
Protein 2g	

Vitamin A 0%	Vitamin C 0%
Calcium 6%	Iron 2%

* Percent Daily Values are based on a 2,000 calorie diet. Your daily values may be higher or lower depending on your calorie needs:

	Calories	2,000	2,500
Total Fat	Less than	65g	60g
Sat Fat	Less than	20g	25g
Cholesterol	Less than	300mg	300mg
Sodium	Less than	2,400mg	2,400mg
Total Carbohydrate		300g	375g
Dietary Fiber		25g	30g

19361

Figure 11.5 The Nutrition Label

(Source: www.nal.usda.gov/fnic/)

In recent years, there has been an increasing interest for consumers to have access to accurate and standardized information about the nutritional content of fresh and ground meat and poultry products. Although not mandatory at this time, the USDA has published guidance for demonstrating voluntary compliance to nutritional labeling requirements. Supermarkets are encouraged to provide consumers accurate nutritional information using "point-of-purchase" product labels, brochures or signage.

Retail establishments that prepare and package product in the store also must provide a label for the consumer. All foods that are packaged in a retail food establishment must be labeled as specified in the law (21 CFR 101 - Food Labeling, and 9 CFR 317 Labeling, Marking Devices, and Containers). The label must include information such as the name of the food, a list of ingredients, and the quantity of ingredients.

Safe Food-Handling Label

A safe food-handling instruction label (Figure 11.6) provides helpful food-handling information for the consumer on all raw or partially pre-cooked (not ready-to-eat) meat and poultry. The label was designed to educate the consumer for storage, preparation, and handling of raw meat and poultry products in the home.

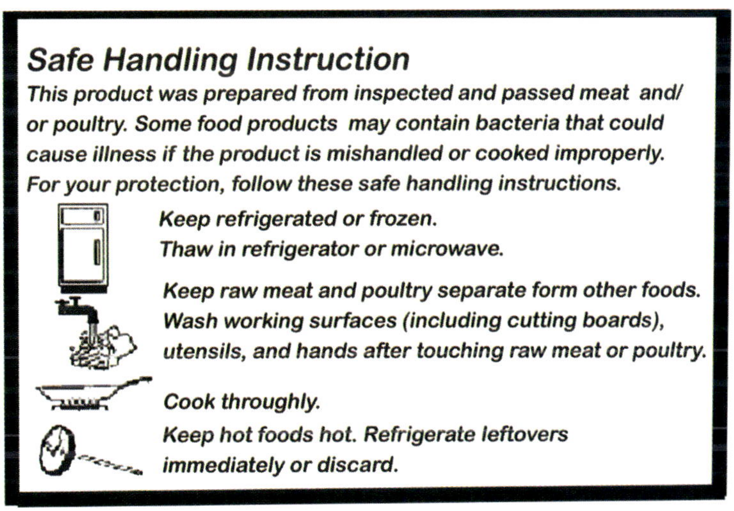

Figure 11.6 USDA Safe Handling Instructions Label (Source: http://www.usda.gov/agency/oc/design/art_symbols.html)

Product Dating Labels

A product dating label is a date stamped on a food package to help a retail food establishment know how long to display the product for sale. It can also help customers understand the time limit to purchase or use the product at its best quality. Date labels are on almost all food products. Date labels are usually found on perishable, potentially hazardous foods such as meat, poultry, eggs, and dairy items.

There are many types of dating labels that can be used. A "sell-by" date tells the retail food establishment how long to display the product for sale. Food products should be removed from the retail shelves immediately after the sell-by date, and customers should buy the product before the date expires. A "best if used by (or before)" date is recommended for the best flavor or quality. If handled properly the product quality will last two to five days past the "sell by" date. A "use by" date is the last date recommended for the use of the product while at peak quality. The date is usually determined by the manufacturer of the product using quality control and shelf life tests. Federal laws regulate date marking of infant formula products.

Recent Initiatives in Food Safety

National Food Safety Initiative

From farm to table, the federal government's National Food Safety Initiative is working to reduce the incidence of foodborne illness by strengthening and improving food safety practices and policies. The initiative seeks to:

- Expand education efforts aimed at improving food handling in homes and retail outlets
- Improve inspections and expand monitoring efforts
- Increase research to develop new and more rapid methods to detect foodborne pathogens and to develop preventive techniques
- Improve intergovernmental communications and coordination of responses to foodborne outbreaks. FDA, CDC, USDA, and EPA will form a new intergovernmental group to improve federal, state, and local responses to outbreaks of foodborne illnesses.

FoodNet

CDC's chief role in the Food Safety Initiative has been to enhance surveillance and investigation of infections usually foodborne. This is the primary mission of FoodNet, a collaborative effort among the Centers for Disease Control and Prevention (CDC), the Food Safety and Inspection Service (FSIS) of the U.S. Department of Agriculture (USDA), the Food and Drug Administration (FDA), and the nine Emerging Infections Program (EIP) sites.

FoodNet was created in 1995 to produce more accurate national estimates of the number of cases and sources of specific foodborne diseases in the United States. The primary objectives of FoodNet are to:

- Determine the frequency and severity of foodborne disease
- Determine the proportion of common foodborne diseases that result from eating specific foods
- Describe the epidemiology of new and emerging bacterial, parasitic, and viral foodborne pathogens.

FoodNet uses an active surveillance program and epidemiological studies to monitor the burden of foodborne diseases over time. FoodNet has also been used to document the effectiveness of new food safety initiatives in improving the safety of our food and the public's health.

PulseNet

PulseNet is a national network of public health laboratories that perform DNA "fingerprinting" on bacteria that may be transmitted by food. The network permits rapid comparison of these "fingerprint" patterns through an electronic database at CDC. Laboratories participating in PulseNet perform DNA "fingerprinting" on disease-causing bacteria isolated from humans and from suspect food, using standardized equipment and methods. Once patterns are generated, they are entered into an electronic database of DNA "fingerprints" at the state or local health department and transmitted to CDC where they are filed in a central computer.

CDC created PulseNet so scientists at public health laboratories throughout the country could rapidly compare the "fingerprint" patterns of bacteria isolated from ill persons and determine whether they are similar. Similar fingerprint patterns suggest the bacteria isolated from ill persons come from a common source, for example, a widely distributed contaminated food product. Strains isolated from food products by regulatory agencies can also be compared with those isolated from ill persons. Identifying these connections can help to detect outbreaks and remove contaminated foods from the marketplace.

The Fight BAC!™ Campaign

The Fight Bac!™ Campaign is an educational program developed by industry, government, and consumer groups to teach customers about safe food handling. The focal point of the campaign is BAC (short for bacteria) a character that is used to make customers aware of the invisible germs that cause foodborne illness (Figure 11.7). Even though customers cannot see, smell, or taste BAC, it and millions more germs like it are in and on food and food-contact surfaces. Fight Bac!™ Campaign uses media and community outreach programs to teach customers how to keep food safe from harmful bacteria through four simple steps: wash hands and surfaces often, prevent cross contamination, cook foods to proper temperature, and refrigerate foods quickly. The partners in the Fight Bac!™ Campaign use newspaper articles, public service announcements (PSAs), printed brochures, and school curricula to teach customers how to lower their risk of foodborne illness.

Figure 11.7 Fight Bac!™ Logo

Consumer Advisory

Food establishments that sell or serve raw or undercooked animal foods or ingredients for human consumption must inform their customers about the increased risk associated with eating these kinds of foods. The food establishment can satisfy the consumer advisory requirements by providing both a disclosure and a reminder. The disclosure is used to make customers aware of the dangers of raw or undercooked foods and of items that either contain or may contain raw or undercooked ingredients. The reminder is a notice about the relationship between thorough cooking and food safety.

Disclosure is satisfied when items are described such as oysters on the half-shell (raw oysters), and raw-egg Caesar salad; or items are asterisked to a footnote that states the items:

- Are served raw, or undercooked or
- Contain (or may contain) raw or undercooked ingredients.

The reminder is satisfied when the items requiring disclosure are asterisked to a footnote that states information about:

- The safety of these items, written information is available upon request
- Eating raw or undercooked meats, poultry, seafood, shellfish, or eggs, which may increase your risk of foodborne illness
- Increased risk of foodborne illness when the above products are eaten if you have certain medical conditions.

Disclosure of raw or undercooked animal-derived foods or ingredients and reminders about the risk of consuming such foods should appear at the point where the customer selects the food. Both the disclosure and the reminder need to accompany the information from which the customer makes a selection. That information could appear in many forms such as display signage, package labels, and consumer brochures.

Consumer advisories may be tailored to be product specific if a food establishment either has a limited menu or offers only certain animal-derived foods in a raw or undercooked, ready-to-eat form. For example, a raw bar serving molluscan shellfish on the half shell, but no other raw or undercooked animal food, could elect to confine its consumer advisory to shellfish.

Summary

The case presented at the beginning of the chapter provides an illustration of a mandatory recall. Whenever a company suspects something may be wrong with one of its products, it may issue a recall to ensure the product is removed from store shelves and that customers who may have already purchased the product are warned not to consume it. Regulatory agencies will frequently participate in recalls to ensure suspect food products are removed from sale as quickly as possible. This is just one example of the partnerships that have been formed between the food industry and regulatory agencies to ensure the safety of our food supply.

Federal, state, and local food safety regulations are important for maintaining a safe food supply. State and local health departments are critical when it comes to monitoring and enforcing each state or local food code. They evaluate food establishments to help ensure delivery of safe food to customers.

Food establishments and regulatory agencies should work together to protect the health of customers and meet their expectations for food quality and safety.

The incident at the beginning of the chapter relates to two major issues: recalls and translated directions. The recall is classified as a Class I – foods that may cause serious adverse health consequences. The person in charge of a retail food establishment must ensure these foods are removed from the shelves immediately.

Problems with formulas in infants are potentially life threatening and must be addressed quickly. Health departments may become involved in removing formulas from homes with infants at risk.

Translation from one language to another is not a simple word-for-word exchange. Meaning and measurement must be checked by a fully qualified linguist and translator who understands the nuances of the different language.

Quiz 11 (Multiple Choice)

1. Which federal agency prepared the model *FDA Food Code*?
    a. Centers for Disease Control and Prevention.
    b. Environmental Protection Agency.
    c. Food and Drug Administration.
    d. Occupational Safety and Health Administration.

2. Which federal agency is responsible for ensuring the safety of domestic meat, poultry, and egg products?
   a. Centers for Disease Control and Prevention.
   b. Environmental Protection Agency.
   c. Food and Drug Administration.
   d. U.S. Department of Agriculture.
3. Which federal agency is responsible for ensuring the safety of all foods except domestic meat, poultry, and egg products?
   a. Centers for Disease Control and Prevention.
   b. Environmental Protection Agency.
   c. Food and Drug Administration.
   d. U.S. Department of Agriculture.
4. Which federal agency is responsible for regulations on safety, noise, and other working conditions?
   a. Centers for Disease Control and Prevention.
   b. Environmental Protection Agency.
   c. Food and Drug Administration.
   d. Occupational Safety and Health Administration.
5. If the FDA determines the ingestion of a food may "cause serious adverse health consequences," a voluntary food recall may be requested. This would be an example of what kind of recall?
   a. Class I.
   b. Class II.
   c. Class III.
   d. Class IV.
6. Which legal document does the FDA abide by to protect the food supply?
   a. Federal Food, Drug, and Cosmetic Act.
   b. 1906 - Federal Meat Inspection Act.
   c. 1957 - Federal Poultry Products Inspection Act.
   d. None of the above.

7. During an inspection of a retail food establishment, the manager should:
   a. Ask to see credentials.
   b. Assist the inspector whenever possible.
   c. Discuss any violations.
   d. All of the above.
8. When a suspected foodborne illness claim is reported to an employee, what should be the first course of action?
   a. Remove all customers from the premises immediately.
   b. Call the health inspector.
   c. Locate the food manager and ensure the suspect food is removed.
   d. Contact CDC.
9. Responsibilities of a food manager include:
   a. Obtaining copies, understanding, and complying with rules and regulations regarding the facility.
   b. Handling retail food establishment self-inspections.
   c. Handling claims of foodborne illness.
   d. All of the above.

Answers to the multiple choice questions are provided in Appendix A.

References/Suggested Reading

Food Safety and Inspection Service, United States Department of Agriculture, Washington, DC. 20250-3700, August 2000, http://www.fsis.gov/oa/pubs/dating.htm

IFT Publication (1992). "Government Regulation of Food Safety: Interaction of Scientific and Societal Forces." *Food Technology.* 46(1).

Labuza, T. P., and Baisier (1992). "The Role of the Federal Government in Food Safety." *Critical Reviews in Science and Nutrition.* 31(3). 167-176.

Potter, N.N., and J. H. Hotchkiss (1995). *Food Science.* Chapman and Hall. New York, NY.

Suggested Web Sites

Conference for Food Protection
www.foodprotect.org

Food and Agriculture Organization
www.fao.org

Food Marketing Institute
www.fmi.org

Gateway to Government Food Safety Information
www.foodsafety.gov

United States Department of Agriculture (USDA)
www.usda.gov

Centers for Disease Control and Prevention (CDC)
www.cdc.gov

United States Environmental Protection Agency (EPA)
www.epa.gov

United States Food and Drug Administration (FDA)
www.fda.gov

USDA Foods Safety and Inspection Service (FSIS)
www.fsis.usda.gov

USDA/FDA Food and Nutrition Information Center
www.nal.usda.gov/fnic/

National Marine Fisheries Services
www.seafood.nmfs.noaa.gov

Occupational Safety and Health Administration (OSHA)
www.osha.gov

Partnerships for Food Safety Education
www.fightbac.org

World Health Organization
www.who.int

APPENDIX A

Answers to Case Studies and Quizzes

Chapter 1

Quiz 1.1	Quiz 1.2
1. b	1. False
2. d	2. False
3. b	3. True
4. d	4. True
5. c	5. True

Chapter 2

Quiz 2

1. d	2. a
3. a	4. c
5. b	6. b
7. c	8. b
9. d	10. a

Chapter 3

Quiz 3

1. a	2. d
3. b	4. b
5. d	6. d
7. d	8. c
9. a	10. b

Chapter 4

Quiz 4

1. c	2. b
3. d	4. c
5. a	6. c
7. b	8. c
9. c	10. d
11. c	12. d
13. c	14. a
15. d	

Chapter 5

Quiz 5

1. c	2. a
3. d	4. d
5. c	6. b
7. d	8. a
9. b	10. d

Chapter 6

Quiz 6

1. c	2. b
3. a	4. a
5. c	6. a
7. d	8. c

Chapter 7

Quiz 7

1. d	2. a
3. b	4. d
5. b	6. a
7. a	8. b
9. a	10. d

Chapter 8

Quiz 8

1. c
2. a
3. c
4. a
5. b

Chapter 9

Quiz 9

1. c	2. d
3. a	4. b
5. c	6. b
7. a	8. c

Chapter 10

Quiz 10

1. a	2. b
3. d	4. b
5. b	6. b
7. a	

Chapter 11

Quiz 11

1. c	2. d
3. c	4. d
5. a	6. a
7. d	8. c
9. d	

Appendix B

Summary of Agents That Cause Foodborne Illness

CAUSATIVE AGENT (*Sporeforming bacteria)	TYPE OF ILLNESS	SYMPTOMS ONSET	COMMON FOODS	PREVENTION
Anasakis spp.	Parasitic infection	Coughing, vomiting 1 hr to 2 weeks	Raw or undercooked seafood, especially bottom-feeding fish	Cook fish to the proper temperature throughout; freeze to meet *2001 FDA Food Code* specification
**Bacillus cereus*	Bacterial intoxication or toxin-mediated infection	1) Diarrhea, abdominal cramps (8 to 16 hrs) 2) Vomiting type, vomiting, diarrhea, abdominal cramps (30 minutes to 6 hrs)	1) Diarrhea type: meats, milk, vegetables 2) Vomiting type: rice, starchy foods; grains and cereals	Properly heat, cool, and reheat foods
Campylobacter jejuni	Bacterial infection	Watery, bloody diarrhea (2 to 5 days)	Raw chicken, raw milk, raw meat	Properly handle and cook foods; avoid cross contamination

CAUSATIVE AGENT (*Sporeforming bacteria)	**TYPE OF ILLNESS**	**SYMPTOMS ONSET**	**COMMON FOODS**	**PREVENTION**
Ciguatoxin	Fish toxin, originating from toxic algae of tropical waters	Vertigo, hot/cold flashes, diarrhea, vomiting (15 minutes to 24 hrs)	Marine finfish, including grouper, barracuda, snappers, jacks, mackerel, triggerfish, reef fish	Purchase fish from a reputable supplier; cooking WILL NOT inactivate the toxin
**Clostridium botulinum*	Bacterial intoxication	Dizziness, double vision, difficulty in breathing and swallowing, headache (12 to 36 hrs)	Improperly canned foods, vacuum packed refrigerated foods, cooked foods in anaerobic mass	Properly heat process anaerobically packed foods; DO NOT use home canned foods
**Clostridium perfringens*	Bacterial toxin-mediated infection	Intense abdominal pains and severe diarrhea (8 to 22 hrs)	Spices, gravy, improperly cooled foods (especially meats and gravy dishes)	Properly cook, cool, and reheat foods
Cryptosporidium parvum	Parasitic infection	Severe watery diarrhea within 1 week of ingestion	Contaminated water, food contaminated by infected food workers	Use potable water supply; practice good personal hygiene and handwashing
Cyclospora cayetanensis	Parasitic infection	Watery and explosive diarrhea, loss of appetite, bloating (1 week)	Water, strawberries, raspberries, and raw vegetables	Good sanitation; reputable supplier

CAUSATIVE AGENT (*Sporeforming bacteria)	TYPE OF ILLNESS	SYMPTOMS ONSET	COMMON FOODS	PREVENTION
Food Allergens	An allergic reaction usually involving the skin, mouth, digestive tract, or airways	Skin: hives, rashes, and itching Mouth: swelling and itching of the lips and tongue Digestive tact: vomiting and diarrhea Airways: difficulty breathing, wheezing	Foods that contain: milk, egg, wheat, nuts and peanuts, fish, and shellfish	Packaged and prepared foods must be properly labeled if they contain common food allergens so sensitive people can avoid them
Giardia lamblia	Parasitic infection	Diarrhea within 1 week of contact	Contaminated water	Potable water supply; good personal hygiene and handwashing
Hepatitis A	Viral infection	Fever, nausea, vomiting, abdominal pain, fatigue, swelling of the liver, jaundice (15 to 50 days)	Foods that are prepared with human contact, contaminated water	Wash hands and practice good personal hygiene; avoid raw seafood
Listeria monocytogenes	Bacterial infection	1) Healthy adult: flu-like symptoms 2) Highly susceptible population: septicemia, meningitis, encephalitis, birth defects (1 day to 3 weeks)	Raw milk, dairy items, raw meats, refrigerated ready-to-eat foods, processed ready-to-eat meats such as hot dogs, raw vegetables, and seafood	Properly store and cook foods; avoid cross contamination; rotate processed refrigerated foods using FIFO to ensure timely use
Mycotoxins	Intoxication	1)Acute onset: hemorrhage, fluid buildup, possible death. 2) Chronic: cancer from small doses over time	Moldy grains: corn, corn products, peanuts, pecans, walnuts, and milk	Purchase food from a reputable supplier; keep grains and nuts dry; protect products from humidity

CAUSATIVE AGENT (*Sporeforming bacteria)	TYPE OF ILLNESS	SYMPTOMS ONSET	COMMON FOODS	PREVENTION
Norwalk virus	Viral infection	Vomiting, diarrhea, abdominal pain, headache, and low grade fever; onset 24 to 48 hrs	Sewage, contaminated water, contaminated salad ingredients, raw clams, oysters	Use potable water; cook all shellfish; handle food properly; meet time temperature guidelines for PHF
Rotavirus	Viral infection	Diarrhea (especially in infants and children), vomiting, low grade fever, 1 to 3 days onset; lasts 4 to 8 days	Sewage, contaminated water, contaminated salad ingredients, raw seafood	Good personal hygiene and handwashing; proper food-handling practices
Salmonella spp.	Bacterial infection	Nausea, fever, vomiting, abdominal cramps, diarrhea (6 to 48 hrs)	Raw meats, raw poultry, eggs, milk, dairy products	Properly cook foods; avoid cross contamination
Scombrotoxin	Seafood toxin originating from histamine-producing bacteria	Dizziness, burning feeling in the mouth, facial rash or hives, peppery taste in mouth, headache, itching, teary eyes, runny nose (1 to 30 minutes)	Tuna, mahi-mahi, bluefish, sardines, mackerel, anchovies, amberjack, abalone	Purchase fish from a reputable supplier; store fish at low temperatures to prevent growth of histamine-producing bacteria; toxin IS NOT inactivated by cooking
Shellfish toxins: PSP, DSP, DAP, NSP	Intoxication	Numbness of lips, tongue, arms, legs, neck; lack of muscle coordination (10 to 60 minutes)	Contaminated mussels, clams, oysters, scallops	Purchase from a reputable supplier

CAUSATIVE AGENT (*Sporeforming bacteria)	TYPE OF ILLNESS	SYMPTOMS ONSET	COMMON FOODS	PREVENTION
Shiga toxin-producing *Escherichia coli*	Bacterial infection or toxin-mediated infection	Bloody diarrhea followed by kidney failure and hemolytic uremic syndrome (HUS) in severe cases (12 to 72 hrs)	Undercooked hamburger, raw milk, unpasteurized apple cider, and lettuce	Practice good food sanitation; handwashing; properly handle and cook foods
Shigella spp.	Bacterial infection	Bacillary dysentery, diarrhea, fever, abdominal cramps, dehydration (1 to 7 days)	Foods that are prepared with human contact: salads, raw vegetables, milk, dairy products, raw poultry, non-potable water, ready-to-eat meat	Wash hands and practice good personal hygiene; properly cook foods
Staphylococcus aureus	Bacterial intoxication	Nausea, vomiting, abdominal cramps, headaches (2 to 6 hrs)	Foods that are prepared with human contact, cooked or processed foods	Wash hands and practice good personal hygiene; cooking WILL NOT inactivate the toxin
Toxoplasma gondii	Parasitic infection	Mild cases of the disease involve swollen lymph glands, fever, headache, and muscle aches. Severe cases may result in damage to the eye or the brain. (10 to 13 days)	Raw meats, raw vegetables and fruit	Good sanitation; reputable supplier; proper cooking
Trichinella spiralis	Parasitic infection from a nematode worm	Nausea, vomiting, diarrhea, sweating, muscle soreness (2 to 28 days)	Primarily undercooked pork products and wild game meats (bear, walrus)	Cook foods to the proper temperature throughout

CAUSATIVE AGENT (*Sporeforming bacteria)	TYPE OF ILLNESS	SYMPTOMS ONSET	COMMON FOODS	PREVENTION
Vibrio spp.	Bacterial infection	Headache, fever, chills, diarrhea, vomiting, severe electrolyte loss, gastroenteritis (2 to 48 hrs)	Raw or improperly cooked fish and shellfish	Practice good sanitation, properly cook foods, avoid serving raw seafood

Appendix C

Employee Health - Disease or Medical Condition Reportable Conditions and Activities

The *2001 FDA Food Code* requires food workers and food employee applicants to report to the person in charge information about their health and activities as they relate to diseases that are transmissible through food. Specifically, an employee must report when he or she:

- Is diagnosed with:
  - *Salmonella* Typhi
  - *Shigella spp.*
  - Shiga toxin-producing *Escherichia coli*
  - Hepatitis A virus.
- Has a symptom caused by gastrointestinal illness or infection such as vomiting, diarrhea, fever, sore throat, or jaundice
- Has a lesion containing pus such as a boil or infected wound that is open and draining
- Had a past illness from:
  - *Salmonella* Typhi within the past three months
  - *Shigella spp.* within the past month
  - Shiga toxin-producing *Escherichia coli* within the past month, or
  - Hepatitis A virus.
- Meets one or more of the following high-risk conditions:
  - Is suspected of causing, or being exposed to, a confirmed disease outbreak caused by *Salmonella* Typhi, *Shigella spp.*, Shiga toxin-producing *Escherichia coli* or Hepatitis A virus where food is implicated in the outbreak
  - Lives in the same household as, and has knowledge about, a person who is diagnosed with a disease caused by *Salmonella* Typhi, *Shigella spp.*, Shiga toxin-producing *Escherichia coli* or Hepatitis A virus

    - Lives in the same household as, and has knowledge about, a person who attends or works in a setting where there is a confirmed disease outbreak caused by *Salmonella* Typhi, *Shigella spp.*, Shiga toxin-producing *Escherichia coli* or Hepatitis A virus.

Exclusions and Restrictions

The person in charge must **exclude** a food employee from a food establishment if the employee:

- Is diagnosed with *Salmonella* Typhi, *Shigella spp.*, Shiga toxin-producing *Escherichia coli,* or Hepatitis A virus
- Works in an establishment that serves a highly susceptible population and he or she:
    - Is experiencing a symptom of acute gastrointestinal illness and meets a high-risk condition as described above
    - Is not experiencing a symptom of acute gastrointestinal illness but has a stool that yields a specimen culture that is positive for *Salmonella* Typhi, *Shigella spp.*, Shiga toxin-producing *Escherichia coli,* or Hepatitis A virus
    - Had a past illness for *Salmonella* Typhi within the last three months or
    - Had a past illness from *Shigella spp.*, Shiga toxin-producing *Escherichia coli* within the last month
    - Is jaundiced
- Is jaundiced and the onset of jaundice occurred within the last seven calendar days.

The person in charge shall **restrict** a food employee from working with exposed food, clean equipment, utensils, and linens; and unwrapped single-service and single-use articles, in a food establishment if the employee is:

- Suffering from symptoms such as diarrhea, fever, vomiting, or sore throat
- Not experiencing a symptom of acute gastrointestinal illness but has a stool that yields a specimen culture that is positive for *Salmonella* Typhi, *Shigella spp.*, Shiga toxin-producing *Escherichia coli* or Hepatitis A virus
- Jaundiced and the onset of jaundice occurred more than seven calendar days before and the establishment does not serve a highly susceptible population.

Removal of Exclusions and Restrictions

The person in charge of a retail food establishment may remove an exclusion if the:

- ▲ Person in charge obtains approval from the regulatory authority
- ▲ Excluded employee provides the person in charge written medical documentation from a physician, or if allowed by law, a nurse practitioner or physician assistant, that specifies:
  - ▼ That the excluded person may work as a food employee in a food establishment, including an establishment that serves a highly susceptible population, because the person is free of the infectious agent of concern
  - ▼ The symptoms experienced by the worker result from a chronic noninfectious condition such as Crohn's disease, irritable bowel syndrome, or ulcerative colitis
  - ▼ The excluded person's stools are free of *Salmonella* Typhi, *Shigella spp.*, or Shiga toxin-producing *Escherichia coli*, whichever is the infectious agent of concern.

The person in charge of a retail food establishment may remove a restriction if the restricted worker:

- ▲ Is free of the symptoms such as diarrhea, fever, vomiting, or sore throat and no foodborne illness occurs that may have been caused by the restricted person
- ▲ Provides written medical documentation from a physician, or if allowed by law, a nurse practitioner or physician assistant, stating that:
  - ▼ The restricted person is free of the infectious agent that is suspected of causing the person's symptoms or causing foodborne illness
  - ▼ Symptoms experienced result from a chronic noninfectious condition such as Crohn's disease, irritable bowel syndrome, or ulcerative colitis
  - ▼ The restricted person's stools are free of *Salmonella* Typhi, *Shigella spp.*, or Shiga toxin-producing *Escherichia coli*, whichever is the infectious agent of concern.

APPENDIX D

Conversion Table for Fahrenheit and Celsius for Common Temperatures Used in Food Establishments

°F	°C		°F	°C
212	100		100	38
200	93		85	30
194	90		75	24
190	88		70	21
180	82		55	13
171	77		45	7
165	74		41	5
160	71		38	3
155	68		36	2
150	66		33	1
145	63		32	0
140	60		30	-1
135	57		29	-2
130	54		0	-18
120	49		-4	-20
110	43		-31	-35

APPENDIX E

Specific Elements of Knowledge Every Retail Food Protection Manager Should Know

I. Identify foodborne illness:

A. Define terms associated with foodborne illness:

1. Foodborne illness.
2. Foodborne outbreak.
3. Foodborne infection.
4. Foodborne intoxication.
5. Disease communicated by food.
6. Foodborne pathogens.

B. Recognize the major microorganisms and toxins that can contaminate food and the problems that can be associated with the contamination:

1. Bacteria.
2. Viruses.
3. Parasites.
4. Fungi.

C. Define and recognize potentially hazardous foods.

D. Define and recognize chemical and physical contamination and illnesses that can be associated with chemical and physical contamination.

E. Define and recognize the major contributing factors for foodborne illness.

F. Recognize how microorganisms cause foodborne illness.

II. Identify time/temperature relationship with foodborne illness.

A. Recognize the relationship between time/temperature and microorganisms (survival, growth, and toxin production) during the following stages:

1. Receiving.
2. Storing.
3. Thawing.
4. Cooking.
5. Holding/displaying.
6. Serving.
7. Cooling.
8. Storing (post-production).
9. Reheating.
10. Transporting.

B. Describe the use of thermometers in monitoring food temperatures:

1. Types of thermometers.
2. Techniques and frequency.
3. Calibration and frequency.

III. Describe the relationship between personal hygiene and food safety.

A. Recognize the association of hand contact and foodborne illness:

1. Hand washing–technique and frequency.
2. Proper use of gloves, including replacement frequency.
3. Minimal hand contact with food.

B. Recognize the association of personal habits and behaviors and foodborne illness:

1. Smoking.
2. Eating and drinking.
3. Wearing clothing that may contaminate food.
4. Personal behaviors, including coughing, sneezing, etc.

C. Recognize the association of the health of a food handler to foodborne illness:

1. Free of systems of communicable disease.
2. Free of infections spread through food on contact.
3. Food protected from contact with open wounds.

D. Recognize how policies, procedures, and management can contribute to and improve food hygiene practices.

IV. Describe methods for preventing food contamination from purchasing to serving.

A. Define terms associated with contamination:

1. Contamination.
2. Adulteration.
3. Damage.
4. Approved source.
5. Sound and safe conditions.

B. Identify potential hazards prior to delivery and during delivery:

1. Approved source.
2. Sound and safe condition.

C. Identify potential hazards and methods to minimize or eliminate hazards after delivery:

1. Personal hygiene.
2. Cross contamination.
    a. Food to food.
    b. Equipment and utensils contamination.
3. Contamination.
    a. Chemical.
    b. Additives.
    c. Physical.
4. Service/display–customer contamination.
5. Storage.
6. Re-service.

V. Identify and apply correct procedures for cleaning and sanitizing equipment and utensils.

A. Define terms associated with:

1. Cleaning.
2. Sanitizing.

B. Apply principles of cleaning and sanitizing.

C. Identify materials, equipment, detergent, sanitizer.

D. Apply appropriate methods of cleaning and sanitizing:

1. Manual warewashing.
2. Mechanical warewashing.
3. Clean-in-place (CIP).

E. Identify frequency of cleaning and sanitizing.

VI. Recognize problems and potential solutions associated with facility, equipment, and layout.

A. Identify facility, design, and construction suitable for food establishments:

1. Refrigeration.
2. Heating and cooling.
3. Floors, walls, and ceilings.
4. Pest control.
5. Lighting.
6. Plumbing.
7. Ventilation.
8. Water supply.
9. Wastewater disposal.
10. Waste disposal.

B. Recognize problems and potential solutions associated with temperature control, preventing cross contamination, housekeeping, and maintenance.

1. Develop and implement self-inspection program.
2. Develop and implement pest control program.
3. Develop and implement cleaning schedules and procedures.
4. Develop and implement equipment and facility maintenance program.

(Developed by the Conference for Food Protection)

(Source: Standards for Accreditation of Food Protection Manager Certification Programs, Annex B. Available at Web site http://www.foodprotect.org)

Glossary

Abrasive cleaners - cleaning compounds containing finely ground minerals used to scour articles; can scratch surfaces.

Acceptable level - within an established range of safety.

Accessible - easy to reach or enter.

Acid - a substance with a pH of less than 7.0.

Additives - natural and man-made substances added to a food for an intended purpose (such as preservatives and colors) or added unintentionally (such as pesticides and lubricants).

Adulterated - the deliberate addition of inferior or cheaper material to a supposedly pure food product in order to stretch out supplies and increase profits.

Aerobe - an organism, especially a bacterium, that requires oxygen to live.

Air curtain - a window or doorway equipped with jets that force air out and across the opening to keep flying insects from entering.

Air-dry - to dry in room air after cleaning, washing, or sanitizing.

Air gap - an unobstructed, open vertical distance through air that separates an outlet of the potable water supply from a potentially contaminated source like a drain.

Alkaline - a substance that has a pH of more than 7.0.

Ambient - a recommended air temperature used to transport or to store perishable foods.

Anaerobe - an organism, especially a bacterium, that does not require oxygen or free oxygen to live.

Anisakiasis - disease caused by the *anisakis* parasite.

Anisakis - roundworm parasite found in fish.

Aseptic packaging - a method in which food is sterilized or commercially sterilized outside the can and then aseptically placed in previously sterilized containers which are then sealed in an aseptic environment. This method may be used for liquid foods such as concentrated milk and soups.

At risk - the term used to describe individuals such as infants, children, pregnant women, and those with weakened immune systems, for whom foodborne illnesses can be very severe, even life threatening.

Backflow - flow of contaminated water into potable supply; caused by back-pressure.

Backroom - a storage area for excess products kept on hand to restock the sales floor as needed.

Back-siphonage - a form of backflow that can occur when pressure in the potable water supply drops below pressure in the flow of contaminated water.

Bacteria - single-celled microscopic organisms.

Bactericide - substance that kills bacteria.

Bacterium - one microorganism.

Bagger - a retail clerk or associate who bags customers' purchases in plastic or paper bags to suit customers.

Bagging - a process of properly, carefully packing a customers' purchases in plastic or paper bags to suit customers.

Bake-off - an in-store baking process using frozen doughs and products to prepare fresh products, i.e., fresh rolls, bread, doughnuts or other pastries.

Bale - a large bundle of cardboard that is recycled.

Baler - a device used to compact and bind corrugated cardboard into bales for recycling.

Bar code - a unique identification code on products, pallets and coupons. The code is read by an electronic scanner for receiving, ordering and inventory control purposes.

Bi-metallic stem thermometer - food thermometer used to measure product temperatures.

Binary fission - the process by which bacteria grow. One cell divides to form two new cells.

Biofilm - a layer of stubborn soil that develops on surfaces that are improperly cleaned or sanitized; bacteria can accumulate and grow on this surface.

Blast chiller - special refrigerated unit that quickly freezes food items.

Bloating - a damaged, swollen processed food can or glass container, which may indicate contamination, a safety hazard.

Bloom - an indication of freshness and quality as beef turns bright red when exposed to oxygen.

Bottle returns - beverage bottles returned to a retailer for recycling.

Botulism - type of food intoxication caused by *C. botulinum.*

Box cutter - a knife-like device with a razor blade used to open boxes.

Braise - to cook meat by browning it in fat, then simmering it in a covered pan with a little liquid.

Breaking down - removing products from a case in order to clean and sanitize it; also, removing component parts of a piece of equipment, such as a slicer or a grinder to clean and sanitize it.

Bulk produce - loose, unpackaged, fresh produce that customers select themselves.

Bulk product - unpackaged, fresh products displayed in bins in large quantities and sold by the piece or the pound, such as grains, candy and snacks.

Calibrate - to determine and verify the scale of a measuring instrument with a standard. Thermometers used in food establishments are commonly calibrated using an ice slush method (32°F or 0°C) or a boiling point method (212°F or 100°C).

Campylobacter jejuni - a microaerophilic non-sporeforming bacterium that causes a foodborne illness.

CAP - controlled atmosphere packaging.

Cardiopulmonary resuscitation (CPR) - first-aid technique to apply to persons who have stopped breathing.

Case, refrigerated - a refrigerated display unit for perishable products, such as dairy products or ice cream.

CBT - computer-based training

Centralized prepackaging - a centralized facility that processes prepackaged perishables and ships the goods to stores.

Channel of distribution - the producers and distributors of products from a farm to the table, a path that includes a grower, producer, manufacturer, broker, wholesaler, store and consumer.

Checker - a front-end employee who rings up, totals and collects for a customer's order. Also known as a cashier.

Checking - the process of recording customer purchases, taking payment, making change, processing coupons, bagging and all other functions inherent to the front-end operation.

Checkout or checkstand - a fixture in a supermarket where customer transactions occur and where the register, scanner, and bags are located. Also known as the front end.

Chemical sanitizers - products used on equipment and utensils after washing and rinsing to reduce the number of disease-causing microbes to safe levels.

Ciguatoxin - a toxin from reef-feeding fish that causes foodborne illness.

Clean - free of visible soil but not necessarily sanitized; surface must be cleaned before it can be sanitized.

Cleaning - removal of visible soil but not necessarily sanitized. A surface must be cleaned before it can be sanitized.

Cleaning agent - a chemical compound formulated to remove soil and dirt.

Clean-in-Place (CIP) - equipment designed to be cleaned without moving, usually large or very heavy.

Code - a systematic collection of regulations, or statutes, and procedures designed to protect the public.

Code date or dating - a date and source code printed on an item to indicate its shelf life. Date codes simplify rotation and help prevent the sale of off-quality or spoiled products. See pull date.

Coffin case or coffin freezer - a waist-high fixture used to display frozen food, with a transparent door or no door for easy access. See upright freezer.

Cold-holding - refers to the safe temperature range (less than 41°F or 5°C) for maintaining foods cold prior to service for consumption.

Cold storage - a facility that stores frozen foods and perishable items that need refrigeration or special handling.

Commingle - to combine shellstock harvested on different days or from different growing areas as identified on the tag or label, or to combine shucked shellfish from containers with different container codes or different shucking dates.

Comminuted - to reduce the size of a food mass by methods including chopping, flaking, grinding, or mincing.

Commissary - a U.S. government-subsidized, nonprofit food store, operating on a military base. Operates on a nonprofit basis and sells products at cost plus a small markup. Also a centralized food preparation facility which distributes prepared product to stores.

Compactor, trash - a device used to crush dry or wet garbage. Often found in many stores in two separate units–one compactor for paper and cardboard and one for all other materials.

Competency based training - a job is analyzed and broken down into steps which must be learned to gain mastery of the task; training should focus on those steps.

Computer-based training (CBT) - a self-paced, interactive, computerized tutorial or training process (see CBT).

Consumer - an end user of any product or service. A shopper, patron or customer. The final link in the chain of product distribution: manufacturing, selling, wholesaling, retailing, consuming.

Contamination - the unintended presence of harmful substances or conditions in food that can cause illness or injury to people who eat the infected food.

Controlled Atmosphere Packaging (CAP) - a product in a low-oxygen, nitrogen-rich wrap that preserves freshness.

Convenience store - a small, easy-access food store with a limited assortment. Many convenience stores also sell fast food and gasoline.

Conventional supermarket - a large, self-service, retail food store with moderate pricing and selection, and annual sales in the $2 to $8 million range. Usually includes a meat, produce, dairy and grocery department.

Cook-chill - a cooking and food preservation process for food service products. Food is cooked, packed, sealed and quick chilled in a plastic pouch and stored at temperatures below 41°F (5°C).

Cooking - the act of providing sufficient heat and time to a given food to affect a change in food texture, aroma, and appearance. More importantly, cooking ensures the destruction of foodborne pathogens inherent to that food.

Cooler - a refrigerated holding unit in a warehouse or store for perishables.

Cooling - the act of reducing the temperature of properly cooked food to 41°F (5°C) or below.

Corrosion resistant - materials which maintain their original surface characteristics under continuous use in food service with normal use of cleaning compounds and sanitizing solutions.

Coving - a curved sealed edge between the floor and wall that makes cleaning easier and inhibits insect harborage.

Critical Control Point (CCP) - means a point or procedure in a specific food system where loss of control may result in an unacceptable health risk.

Critical limit - the maximum or minimum value to which a physical, biological, or chemical parameter must be controlled at a critical control point to minimize the risk that the identified food safety hazard may occur.

Cross-connection - any physical link through which contaminants from drains, sewers, or waste pipes can enter a potable water supply.

Cross contamination - transfer of harmful organisms between items by direct or indirect contact.

Cryovac - a proprietary term for vacuum packaging material, which has entered the language to mean all vacuum packaging, like "Kleenex" or "Band-aid."

Cumulative - increasing in effect by successive additions. For example, hot food must be cooled from 140°F (60°C) to 70°F (21°C) within two hours; it must reach 41°F (5°C) within an additional four hours to prevent bacterial growth. Therefore, the cumulative time the food is in this danger zone equals a total of six hours or less.

Cutting - opening or sampling a product to evaluate its appearance, flavor, quality and/or consistency.

Dairy/deli case extender - an insulated container display attached to a refrigerated case that extends into an aisle to stimulate impulse buys.

Danger zone - temperatures between 41°F (5°C) and 140°F (60°C).

Date code or coding - a "sell by" date stamped on a product to ensure freshness. The date assists with quality control (first-in, first-out) and proper rotation. It may also apply to affixing a "sell by" or "pull by" date on merchandise which is on display (as in the Bakery Department).

Disinfectant - destroys harmful bacteria.

Dead man's switch - activates equipment when depressed and stops if pressure is relieved.

Deli-bake - a combination in-store bakery and deli department where equipment, floor space and labor are shared, usually under common supervision.

Delicatessen - an in-store department with cooked foods, salads, cold cuts and cheeses, etc.

Demo or demonstration - a product promotion in a store with samples to eat and cooking-tip handouts or coupons.

Detergents - cleaning agent which contains surfactants used with water to break down soil to make it easier to remove.

Deviation - to diverge; to go in different directions.

Digital thermometer - a battery-powered thermometer that reveals temperature in a digital numerical display.

Dock - an area to receive, load and unload shipments.

Dressed fish - a whole scaled, cleaned fish, sold with or without the head.

Dry grocery - nonperishable grocery products.

Dry grocery non-foods - products that are not food, such as paper products, detergents, or pet items.

Easy-to-clean - materials and design which facilitate cleaning.

Easy-to-move - on wheels, raised on legs, or otherwise designed to facilitate cleaning.

Employee - person working in or for a food establishment who engages in food preparation, service, or other assigned activity.

Equipment - the appliances: stoves, ovens, etc.; and storage containers, such as refrigerated units used in food establishments.

Ethnic foods - products that a particular ethnic (racial, national, religious, or cultural) group favor such as Mexican, Chinese, or Kosher.

Evaluation procedures - systematically checking progress to determine if goals have been met.

Exotic produce - fruits and vegetables not grown in North America and considered exotic such as kiwi fruit and ugli fruit.

Expiration date - a manufacturer's "sell by" date stamped on products to indicate shelf life.

Facultative anaerobe - an organism that can grow with or without free oxygen.

FATTOM - an acronym used to indicate the six conditions bacteria need for growth. These conditions are Food, Acid, Temperature, Time, Oxygen, and Moisture.

FIFO - acronym for first-in, first-out, used to describe stock rotation procedures of using older products first.

FMI - Food Marketing Institute - an international trade association of independent grocers, chain stores, and wholesalers.

Food Allergen - a substance in food that causes the human immune system to produce chemicals and histamines in order to protect the body. These chemicals produce allergic symptoms that affect the respiratory system, gastrointestinal tract, skin, or cardiovascular system.

Foodborne disease outbreak - an incident in which two or more people experience a similar illness after ingesting a common food, and in which epidemiological analysis identifies the food as the source of the illness.

Foodborne illness - an illness caused by the consumption of a contaminated food.

Food - contact surface - any surface of equipment or any utensils which food normally touches.

Food establishment - an operation that stores, prepares, packages, serves, vends, or otherwise provides food for human consumption such as a restaurant, food market, institutional feeding location, or vending location.

Footcandle - unit of lighting equal to the illumination one foot from a uniform light source.

Forklift - a vehicle with projecting prongs that slide under a pallet to move merchandise in a warehouse or store.

Fresh - just picked, gathered, produced, live or unprocessed, not stale, food. A term associated with perimeter departments, including produce, deli, bakery, or floral; also, unfrozen.

Fungi - a group of microorganisms that includes mold and yeasts.

Garbage - wet waste matter, usually food product, that cannot be recycled.

Garnish - a decoration on salads, such as sprigs of watercress, lettuce or other colorful items.

Gastroenteritis - an inflammation of the linings of stomach and intestines which can cause nausea, vomiting, abdominal cramping, and diarrhea.

General merchandise (GM) - products other than food sold in supermarkets and require special buying, warehousing and servicing. GM classes are: hardiness, softness, reading/writing lines, health and beauty care (HBC) and services.

Generally recognized as safe (GRAS) - a food safety FDA term that indicates all ingredients are approved for human consumption. (See GRAS Substances.)

Germicide - a substance which kills harmful microbes and germs.

Germs - general term for microorganisms, including bacteria and viruses.

Grade - a food industry classification system or standard that indicates a quality level, such as Grade A, Prime, or Extra Fancy.

Grade standards - primarily standards of quality to help producers, wholesalers, retailers, and patrons in marketing and purchasing food products. The grade standards are not aimed at protecting the health of the consumer but rather at ensuring value received according to uniform quality standards.

GRAS substances - GRAS stands for generally recognized as safe. These are substances added to foods that have been shown to be safe based on a long history of common usage in food.

Grocery store - a retail store that sells a variety of food products, including some perishable items and general merchandise.

Grocery wholesaler - a middleman who buys food and supplies from manufacturers to resell them in smaller quantities to retailers; cooperatives and voluntaries are the two major types.

HACCP - Hazard Analysis Critical Control Point.

Hand washing - the proper cleaning of hands with soap and warm water to remove dirt, filth, and disease germs.

Harborage - shelter for pests.

Hazard - a biological, chemical, or physical agent that may cause an unacceptable consumer health risk.

Hazard analysis - to identify hazards (problems) that might be introduced into food by unsafe practices or the intended use of the product.

Hazard Analysis Critical Control Point (HACCP) - a federal guideline to ensure safe food handling and preparation from receiving to point of sale. (See HACCP.)

Heat-and-eat - a precooked food that requires heating before consumption.

Heimlich maneuver - method used to expel foreign body caught in someone's throat.

Hermetic packaging - a container completely sealed against the entry of bacteria, molds, yeasts, and filth as long as it remains intact.

Highly susceptible population - persons who are more likely than other people in the general population to experience foodborne disease because they are immunocompromised, preschool age children, or older adults.

Home meal replacement - foods prepared in a store and consumed at home or in-store which require little or no preparation on the part of the consumer.

Hot-holding - refers to the safe temperature range of 140°F (60°C) and above to maintain properly cooked foods hot until served.

Housekeeping - operational procedures to ensure cleanliness, safety, sanitation and maintenance for a store or warehouse.

Immunosuppressed - an individual who is susceptible to illness because his or her immune system is not working properly or is weak.

Impermeable - does not permit passage, especially of fluids.

Individually quick frozen (IQF) - a food processing technique that freezes products in the final stage of processing. It is then wrapped and packaged for shipment.

Infection - illness caused by eating food that contains living disease-causing microorganisms.

Infestation - presence of a large number of pests.

Ingestion - the process of eating and digesting food.

Inherent - being an essential part of something.

In-store bakery - a bakery department producing baked goods in the store using one of two baking methods: scratch using ready-made mixes and frozen dough, making products according to a formula or bake-off.

In-store demonstrators - an employee or temporary person conducting tastings, samplings, and product preparations to encourage impulse buys from customers.

Insulated containers - containers used to hold hot or cold products to ensure that they will be in good condition when they get to the customer's home.

Integrated Pest Management (IPM) - a system of preventive and control measures used to control or eliminate pest infestations in food establishments.

Intoxication - illness caused by eating food that contains a harmful chemical or toxin.

Irradiation - a food preservation process that utilizes radiation to control bacteria growth and increase shelf life.

Kick plate (base) - a metal sheet, usually at the bottom of doors, for protection purposes.

Kitchenware - utensils used to prepare food.

Larva - immature stage of development of insects and parasites.

Leftovers - any food that is prepared for a particular meal and is held over for service at a future meal.

Low temp - A refrigerator that holds products at a below-freezing temperature 32°F (0°C)) or less.

Manager - the individual present at a food service establishment who supervises employees who are responsible for the storage, preparation, display, and service of food to the public.

MAP - modified atmosphere packaging.

Measuring device - usually a thermometer that registers the temperature of products and water used for sanitizing.

Media - newspapers, magazines, radio, and television.

Mesophile - microorganisms that grow best at moderate temperatures.

Microbe - a microorganism that can cause disease. Bacteria, molds and yeast that can grow on various food and equipment surfaces; the main cause of discoloration in meat and food poisoning.

Microorganism - bacteria, viruses, molds, and other tiny organisms that are too small to be seen with the naked eye. The organisms are also referred to as microbes because they cannot be seen without the aid of a microscope.

Misbranding - falsely or misleadingly packaged or labeled food; may contain ingredients not included on label or does not meet national standards for that food.

Modified Atmosphere Packaging (MAP) - a food processing technique where foods are placed in a flexible container and the air is removed from the package. Gases may be added to help preserve the food. (See MAP.)

Mold - any of various fungi that spoils food and has a fuzzy appearance.

Monitoring procedures - a defined method of checking foods during receiving, storage, preparation, holding, and serving processes.

Nonfood-contact surface - any area not designed to touch food.

Nutrition labeling - an accurate list of ingredients printed on food, beverage and drug labels.

Occupational Safety and Health Administration (OSHA) - a federal agency that sets workplace safety standards and inspects facilities for safe working conditions.

OJT - on-the-job training.

Onset time - the period between eating a contaminated food and developing symptoms of a foodborne illness.

On-the-job-training - an instructional method to teach employees skills as they work. (See OJT.)

Open dating - a date stamped or printed on the label of perishable items to indicate a pull date (a date by which the item must be sold or removed from the shelf) or pack date (the date the item was packaged). Clear, readable dates that are printed on labels.

Organically grown - an imprecise term that means a grower did not use chemicals or that a processor did not use preservatives in the product.

Ovenable - a food ready to be heated, either in an oven or microwave.

Overstock - an excessive amount of product purchased in anticipation of increased sales volume.

Over-wrap - to wrap a plastic container in cellophane to prevent tampering.

Pack date - the date on which a product was made or packaged for sale.

Palatable - food that has an acceptable taste and flavor.

Pallet - a standard-sized base for assembling, sorting, stacking, handling and transporting goods as a unit. The industry standard is GPC-spec-4-way entry, 48 inches x 40 inches hardwood pallets.

Parasite - an animal or plant that lives in or on another and from whose body it obtains nourishment.

Parts per million (ppm) - unit of measure for water hardness and chemical sanitizing solution concentrations.

Pasteurization - a low heat treatment used to destroy disease-causing organisms and/or extend the shelf life of a product by destroying organisms and enzymes that cause spoilage.

Pathogenic - capable of causing disease; harmful; any disease-causing agent.

Perishables - foods requiring refrigeration or special handling because they spoil easily, such as meat, seafood, produce, deli, bakery and dairy.

Personal hygiene - health habits including bathing, washing hair, wearing clean clothing, and proper hand washing.

Pest control operator - licensed individual or certified technician who provides pest control services.

pH - the symbol that describes the acidity or alkalinity of a substance, such as food.

Physical hazard - particles or fragments of items not supposed to be in foods.

Point-of-purchase - the locations within a retail store where a customer purchases products.

Point of sale (POS) - the place in a retail store where products are scanned through the register system, data is collected, and sales are tendered. POS also describes sales data generated by checkout scanners.

Potable water - water that is safe to drink.

Potentially hazardous food - a food that is natural or man-made and is in a form capable of supporting the rapid and progressive growth of infectious and toxin-producing microorganisms. The foods usually have high protein and moisture contents and low acidity.

Pounds per square inch (psi) - amount of pressure per square inch.

Premises - physical environment of the food establishment; grounds, interior, and exterior of building(s).

Preventive measures - procedures that keep foods from becoming contaminated.

Proof box - a piece of equipment, in which heat and humidity are controlled in order for dough to rise in preparation for baking.

Psychrophiles - microorganisms that grow best at cold temperatures.

Pull date - The date by which a product must be either sold or pulled from a shelf.

Quaternary ammonium - a chemical sanitizing compound that is relatively safe for skin contact and is generally noncorrosive; effective in both acid and alkaline solutions.

Reach-in case - a refrigerated display case with a self-service door used for perishable products.

Ready-to-eat foods - products that are in a form that is edible without washing, cooking, or additional preparation by the retail food establishment or the patron.

Receiver - an authorized associate of a warehouse or retail store who receives and checks deliveries for condition and accurate amount. The first handler of the delivery receipt or invoice.

Receiving - a door or dock of a warehouse or store designated for receiving merchandise from a supplier. The procedure for physically and legally accepting a shipment of product.

Receiving clerk - see receiver.

Receiving log - the record or listing of products received with appropriate entries.

Reclaimed goods - Unsalable product at the time of delivery that is returned to a wholesaler/vendor for reclamation.

Reconstitute - to combine dehydrated foods with water or other liquids to bring back to its original state; example, addition of water to powdered milk.

Refuse - trash, rubbish, waste.

Reheating - the act of providing sufficient heat (at least 165°F or 74°C) within a two-hour time period to ensure the destruction of any foodborne pathogens that may be present in that cooked and cooled food.

Returns - unsold, damaged, or defective merchandise sent to a supplier or distributor for credit or refund.

Rework - perishables: To crisp or trim a product that looks case-worn. Grocery: To re-affix labels. To refine a category or shelf set.

Re-wraps - products that are removed, reconditioned (if salable) and displayed with limited sell-by dates.

Risk - the chance of injury, damage, or loss.

Rotation - a shelf-stocking procedure that ensures first-in, first-out by pulling older stock forward and placing newer stock at the back during restocking.

Rotisserie - a rotating grill with an electrically turned spit that cooks meats.

Safety cutter - a case cutter used to open cases of product.

Salvage - product containers/shippers (bales, pallets, containers) that must be returned or recycled to defray operational costs.

Sanitary - healthful and hygienic. The number of harmful microorganisms and other contaminants have been reduced to safe levels.

Sanitation - maintenance of conditions which are clean and promote good health.

Sanitizer - approved substance or method to use when sanitizing.

Sanitizing - application of an agent that reduces microbes to safe levels.

Sealed - closed tightly.

Segregation - locating general merchandise products (GM) in a well-defined area of a store rather than in aisles next to or across from food products.

Selectivity - chemical sanitizers, especially quats, may kill only certain organisms and not others.

Sensitive ingredient - an ingredient that is prone to support bacterial growth.

Shelf life - Time period a product can be expected to maintain maximum quality and freshness.

Shelf stable - A processed food product that remains safe to eat without refrigeration.

Shellfish - an aquatic animal, i.e., clams, oysters, mussels, crabs, lobsters, shrimp, squid and octopus which has a shell; crustaceans or mollusks.

Shellstock - raw, in-shell molluscan shellfish.

Shiga toxin-producing *Escherichia coli* - any *E. coli* capable of producing Shiga toxins (also called verocytotoxins or "Shiga-like" toxins). This includes, but is not limited to, *E. coli* reported as serotype O157:H7, O157:NM, and O157:H-.

Shucking - a process of opening shellfish, such as oysters, clams, mussels, etc.

Silicate - salt or ester derived from silica; a hard, glassy material found in sand.

Single-use articles - items intended for one use and then discarded, such as paper cups or plastic eating utensils.

Slacking - the process used to moderate the temperature of a food such as allowing a food to gradually increase from -10°F (-23°C) to 25°F (-4°C) in preparation for deep-fat frying or to facilitate even heat penetration during the cooking or reheating of a previously block-frozen flood such as spinach.

Sneeze guard - a clear, solid barrier which partially covers food in self-service areas to keep customers from coughing, sneezing, or projecting droplets of saliva directly onto food.

Soil - dirt and filth.

Sous-vide - a European food-packaging technique where a prepared product is placed in individual pouches, cooked under a vacuum and quickly chilled. Products are frozen or refrigerated until used.

Splash contact surfaces - areas that are easy to clean and designed to catch splashed substances in work areas.

Spoilage - significant food deterioration, usually caused by bacteria and enzymes, that produces a noticeable change in the taste, odor, or appearance of the product.

Spoils - goods that cannot be sold for which a retailer receives a credit from a supplier. Also called stales.

Spore - the inactive or dormant state of some rod-shaped bacteria.

Standard of identity - Food and Drug Administration (FDA) standards for food composition.

Standard operating procedures - a comprehensive book of a company's policies and procedures. Also called SOPs.

Stationary equipment - equipment that is permanently fastened to the floor, table, or counter top.

Sulfites - preservatives used to maintain freshness and color of fresh fruits and vegetables; subject to state regulations.

Supermarket - A conventional grocery store, but not a warehouse club or mass merchant, with annual sales of $2 million or more per store.

Supplier - a generic term for wholesalers, who sell to and supply retailers directly and indirectly, i.e., manufacturer, vendor, broker, reseller.

Surfactant - chemical agent in detergent that reduces the surface tension, allowing the detergent to penetrate and soak soil loose; wetting agent.

Swells - unsalable items with expanded containers or lids signifying faulty food handling, processing or sealing.

Tableware - plates, cups, bowls, etc.

Task analysis - examining a job task to determine what it takes to do the job.

Temperature abuse - allowing foods to remain in the temperature danger zone (41° and 140°F, 5° and 60°C) for an unacceptable period of time.

Temperature danger zone - temperatures between 41° and 140°F (5° and 60°C) at which bacteria grow best.

Test kit - device that accurately measures the concentration of sanitizing solutions to ensure they are at proper levels.

Thermometer - a device that measures temperatures.

Thermophiles - microorganisms that grow best at hot temperatures.

Time and temperature monitoring (TTM) devices - label-sized, chemical-filled packets that are attached to shipping cartons and indicate when the cartons have been exposed to fluctuating temperatures.

Toxin - a poisonous substance produced by microorganisms, plants, and animals and which causes various diseases.

Toxin-mediated infection - illness caused by eating a food that contains harmful microorganisms that produce a toxin once they get inside the human intestinal tract.

Trainee - an employee participating in a company-sponsored training program.

United States Department of Agriculture (USDA) - a federal agency that oversees food production and inspection. The USDA establishes grade standards for commodities, conducts agricultural research and makes results available, administers food programs (i.e., food stamps), and distributes food.

United States Department of Commerce (USDC) - a federal agency that oversees trade and competition. The USDC establishes grade standards for seafood commodities, conducts agricultural research, and makes results available.

United States grade stamp - signifies that a product is clean, safe and wholesome, and has been produced in an acceptable establishment, with the appropriate equipment, under the supervision of federal inspectors. It also indicates the product is of a specific grade.

Unsalables - products unworthy of sale, e.g., damaged, out of date, spoiled.

Upright freezer - an upright refrigerated display unit with doors used for merchandising frozen foods.

USDA Grade - United States Department of Agriculture grades that relate to a specified quality of product. Grade denotes quality and USDA denotes product inspected for wholesomeness.

Utensil - a food-contact implement or container used in the storage, preparation, transportation, dispensing, sale, or service of food, such as kitchenware or tableware that is multi-use, single-service, or single-use; gloves used in contact with food; temperature sensing probes of food temperature-measuring devices; and probe-type price or identification tags used in contact with food.

Vacuum breaker - designed for use under a continuous supply of pressure. Spring loaded device to operate after extended periods of hydrostatic pressure.

Vacuum Packaging - a packaging process in which air is removed from a package as it is sealed.

Vegetative state - the active state of a bacterium where the cell takes in nourishment, grows, and produces wastes.

Ventilation - air circulation that removes smoke, odors, moisture, and grease-laden vapors from a room and replaces them with fresh air.

Verification - to prove to be true by evidence, usually a record of time and temperatures of food from receiving to serving or vending.

Viruses - any of a group of infectious microorganisms that reproduce only in living cells. They cause diseases such as mumps and Hepatitis A and can be transmitted through food.

Warewashing - the cleaning and sanitizing of utensils and food-contact surfaces of equipment.

Water activity (A_w) - a measure of the free moisture in a food. Pure water has a water activity of 1.0, and potentially hazardous foods have a water activity of 0.85 and higher.

Waxing - applying an edible wax to some fruits and vegetables to help maintain a fresh, bright appearance and to preserve product quality.

Wetting agent - a substance which breaks down the soil to allow water and soap or detergent to liquefy and remove dirt and grease.

Whole-muscle, intact beef - whole-muscle beef that is not injected, mechanically tenderized, reconstructed, or scored and marinated, from which beef steaks may be cut.

Wholesome - something that is favorable to or promotes health.

Yeast - type of fungus that is not known to cause illness when present in foods but can cause damage to food products and will change taste; useful in making products such as bread and beer.

INDEX

A

accident prevention 14–15, 302–313
air curtain
  figure 284
air gap
  figure 273
Anisakis spp. 56

B

Bacillus cereus 44
bacteria
  examples of 41
  foods that support growth of
    figure 34
  growth and acidity
    figure 35
  growth curve
    figure 32
  growth of
    figure 37
  growth temperature
    figure 36
  growth time 37
    figure 37
  multiply
    figure 33
  spores
    figure 42
binary fission
  figure 31
biological hazards
  figure 41
boiling point method 84

C

calibrate
  figure 84
cans
  defective
    figure 117
Cardiopulmonary Resuscitation (CPR)
  figure 310
CDC 5, 6
Centers for Disease Control and Prevention 5
certified food protection manager 18
CFP 17
chemical
  storage 135
    figure 135
Ciguatoxin 64
classifications of foodborne illness 26
clean
  figure 14
cleaning 13–14, 216
  figure 14
Clostridium botulinum 46
Cockroaches 286
cockroaches
  case study 261
  figure 286
cold food recommendations for 88–90
  figure 90
cold-holding 150
conference for food protection 17
contaminated equipment 77
contamination 9
  sources of
    figure 134
controlling temperatures 87
convection oven
  figure 184
convenience store 4

cooking
convection oven figure 184
deck oven figure 183
foods 145–147
figure 146
microwave oven figure 184
times and temperatures
figure 146
cooling
foods danger of 88–90
figure 90
foods guidelines for 147–149
figure 148, 149
methods of 147
figure 148, 149
crisis management 14–15
cross contamination 11
figure 100, 101
prevention 101–102
Cryptosporidium parvum 58

D

dairy products 124–125
design 172
dishwashers
single-tank, stationary-rack
figure 202, 236
dishwashing
machine temperature
figure 238
manual
figure 200
mechanical
figure 202, 236

E

education 15
eggs
grading of 123
figure 123
inspection of 123
figure 123
receiving of 123
equipment
figure 180
refrigeration 185–190

F

facilities 172
facility
design 172
planning and design 13
FDA Food Code 17
fire
extinguishers
figure 311
flies
control of
figure 284
figure 283
flow of food 113
fluid milk 124
FMI 5
food
code preface 16
establishment
figure 3
establishment exterior condition of
industry functions of 2–4
flow of
figure 114
poisoning 24–27
storage types of 130–135
temperatures 78
temperatures measurement of 81–83, 113
temperatures recommended safe
figure 87
trends 6–8
Food and Drug Administration (FDA) 376
food code 16
food establishment 2
food flow 11
Food Marketing Institute 5
food protection manager certification 17–19

food safety
- role of government 16
- role of industry 17

foodborne disease outbreak 8
foodborne illness 5, 6, 8, 25
- factors of
  - figure 77
- prevention of
  - figure 7
- types of
  - figure 25

foodborne intoxications 26
frozen foods storage of 129
fruits and vegetables receiving of 127

G

game animals 121
garbage
- refus station
  - figure 279

germs 11
gloves (disposable) proper use of
- figure 96

H

HACCP 12
handwashing 77
- facilities 270–271
  - figure 270
- guidelines
  - figure 92
- proper technique for
  - figure 93
- station
  - figure 94, 270

handwashing and good personal hygiene 91
handwashing guidelines 92
handwashing station 94
Hazard Analysis Critical Control Point 12
heat sensitive labels
- figure 229

Hepatitis A 54
home meal replacement 162
hood
- figure 207

hot-holding 150

I

ice machines 193
ice point method of calibration 84
- figure 84

improper cooling of foods 89
improper holding temperature 77
infection 26
intoxication 26

L

layout 172
lighting
- figure 206

M

manual washing 200
material safety data sheets
- figure 253

measuring food temperature 84
microbes 11
microorganisms 11
- figure 11

milk
- fluid 124
- products 124

mobile food facilities 159
- figure 159

Modified Atmosphere Packaging (MAP)
- figure 118

Mycotoxins 66

N

Norwalk virus 55
nutrition label 376

P

packaged foods 116

parasites
figure 41
freezing guidelines
firgure 139
personal health 98
personal hygiene 97–98
clothing and apparel 96–97
disposable gloves
figure 96
personal health reportable diseases 98
recommendations for 91–98
poor personal hygiene 77
potentially hazardous foods 12
figure 40
poultry 121
grading of 121–122
figure 122
inspection of
figure 122

R

rats
figure 289, 290
receiving 113–114
frozen foods 129
packaged foods 116–117
red meats 119–120
requirements of 114
receiving of
eggs 123
red meats
grade of
figure 120
inspection of
figure 120
receiving of 119–120
spoilage of 119–120
refrigerator
maximum load line in refrigerated display case
figure 186
standard upright
figure 187
under-the-counter
figure 187
reheating foods
recommendations 88
requirements for 150
retail food establishments 4
rodents
droppings
figure 290
Rotavirus 55

S

sanitary 14
figure 14
sanitation 14
sanitizing 216
heat 227–230
figure 229
Scombrotoxin 64
self-service bar keeping food safe at 156–157
figure 156, 157
serving 154
shellfish
tags
figure 127
toxins 65
Shiga toxin-producing Escherichia coli 19, 48
slicer
figure 192
soft serve machine
figure 247
Staphylococcus aureus 52
storage 113
cold 129
dry
figure 133

T

temperature 128, 138
  abuse 138
    prevention 87–88
      figure 87, 90, 140
  dishwashing machines
    figure 238
temperature danger zone 79
temperature measuring device
  bi-metal thermometer 81
  calibration of 83–84
    figure 84
  cleaning and sanitizing of 83
  T-stick
    figure 229
  types of 81–83
temperature measuring divice
  maximum registering thermometers
    figure 230
temporary food facilities 159–161
  figure 159
thawing 140
time and temperature abuse 78
toxin-mediated infection 26
Toxoplasma gondii 59
training 15
traps 293
Trichinella spiralis 60

V

ventilation
  figure 207
Vibrio spp. 53
viruses
  figure 41

W

water activity
  figure 39